Ultra Wideband Antennas

DESIGN, METHODOLOGIES, AND PERFORMANCE

Giselle M. Galvan-Tejada
COMMUNICATIONS SECTION
DEPARTMENT OF ELECTRICAL ENGINEERING
CINVESTAV-IPN
MEXICO CITY, MEXICO

Marco A. Peyrot-Solis
INIDETAM
MEXICAN NAVY
VERACRUZ, MEXICO

Hildeberto Jardón-Aguilar
COMMUNICATIONS SECTION
DEPARTMENT OF ELECTRICAL ENGINEERING
CINVESTAV-IPN
MEXICO CITY, MEXICO

CRC Press
Taylor & Francis Group
Boca Raton London New York

CRC Press is an imprint of the
Taylor & Francis Group, an **informa** business

CRC Press
Taylor & Francis Group
6000 Broken Sound Parkway NW, Suite 300
Boca Raton, FL 33487-2742

First issued in paperback 2017

© 2015 by Taylor & Francis Group, LLC
CRC Press is an imprint of Taylor & Francis Group, an Informa business

No claim to original U.S. Government works

ISBN-13: 978-1-4822-0650-0 (hbk)
ISBN-13: 978-1-138-89381-8 (pbk)

Library of Congress Cataloging-in-Publication Data

Galvan-Tejada, Giselle M., author.
  Ultra wideband antennas : design, methodologies, and performance / authors, Giselle M. Galvan-Tejada, Marco Antonio Peyrot-Solis, Hildeberto Jardon Aguilar.
    pages cm
  Includes bibliographical references and index.
  ISBN 978-1-4822-0650-0 (hardcover : alk. paper) 1. Ultra-wideband antennas. I. Peyrot-Solis, Marco Antonio, author. II. Aguilar, Hildeberto Jard?n, author. III. Title.

  TK7871.67.U45G38 2015
  621.3841'35--dc23   2015026062

Visit the Taylor & Francis Web site at
http://www.taylorandfrancis.com

and the CRC Press Web site at
http://www.crcpress.com

To my beloved husband, Aldo Gustavo and young children Gustavo Stuart and Minerva Montserrat, for their love, support and encouragement all the time. You are always in my mind and heart.

To my mom for giving me the warm space of her being and my dad for giving me the blood of an engineer.

To all my brothers and sisters who were always supportive during the progress of this book.

Giselle

This book is dedicated to my wife Cynthia, and my son and daughter Marco Antonio Cesar and Cynthia Carolina for their patience and support through the long nights when this work was completed.

Marco Antonio

Contents

List of Figures

List of Tables

Preface

Over the past decades ultra wideband antennas have attracted the attention of the scientific community due to their wide variety of applications such as body area networks, radar, imaging, spectrum monitoring, electronic warfare, and wireless sensor networks, among others. As a result, a vast quantity of work presenting diverse designs has been reported around the world. In spite of this, some other possibilities have to be explored in the UWB antenna's design in such a way that current challenges may be solved.

Thus, this book was conceived as reference material for the development of UWB antennas. Different aspects are considered in this text, from recent proposals on ultra wideband antennas reported in diverse forums, theory specific for these radiators, up to guidelines for the design of omnidirectional and directional UWB antennas.

According to current tendencies, two types of antennas are identified based on their structures, planarized and planar, to distinguish between those flat radiators embedded and non-embedded on the ground plane. An important concept used here is the solid-planar equivalence, which allows that flat structures can be implemented instead of volumetric antennas. This principle is vital for the developments of UWB antennas on portable equipment and for the recent body area networks, where small low profile radiators are intended to integrate to wearable devices.

Time domain signal analysis for UWB antennas, from which the distortion phenomenon can be modeled through group delay and phase linearity, is addressed as well. It is one of the main differences with traditional narrowband or wideband antennas where transient response has not been considered. Thus, in particular, some important quantities associated with the impulse response of UWB antennas are reviewed.

Design methodologies for omnidirectional and directional antennas are described, and the dependence on their operation as a function of distinct factors (ground plane, beveling on radiator, height/wide proportion, reflector, etc.) is examined in depth. In all cases, three objectives are considered: Impedance matching, phase linearity, and variations of the shape of the radiation pattern. Performance comparisons among different reported UWB designs are discussed.

Some current tendencies and unresolved problems in the field of UWB antennas are also noted. The book ends with a brief exposition of numerical techniques for electromagnetics, including the generalities of the classical Finite Differences Method, Finite Element Method, and Method of Moments.

Certain antenna models are taken to illustrate particular conceptual aspects of these methods.

Outline

In summary, the main points addressed in this book are

- An outline of recent developments on UWB antennas

- Developed theory for UWB antennas in frequency and time domains

- Design methodologies for omnidirectional and directional UWB antennas

- Performance comparisons of different UWB antennas

- A vision of tendencies and unresolved problems to date

- An exposition of numerical methods for electromagnetics oriented to antennas

Regarding simulations, it is worth mentioning that CST Microwave Studio was the platform used through this book.

For product information, please contact:
CST-Computer Simulation Technology AG
Bad Nauheimer Strasse 19
64289 Darmstadt, Germany
Tel: +49 6151 7303 0
Fax: +49 6151 7303 100
Email: info@cst.com
Web: www.cst.com

Plots presented in distinct sections of chapters were generated using MATLAB®.

For product information, please contact:
The MathWorks, Inc.
3 Apple Hill Drive
Natick, MA 01760-2098 USA
Tel: 508-647-7000
Fax: 508-647-7001
E-mail: info@mathworks.com
Web: www.mathworks.com

Acknowledgments

The authors are deeply grateful to MSc Ruben Flores-Leal for his support in simulating many antenna structures, and Mrs. Eva Ojeda-Sanchez for her help in drawing a vast quantity of figures through the whole manuscript. Seeing our ideas transformed into electronic images was possible thanks to the amazing tasks of both appreciated persons, who were always in total support of this book. This work was also enabled by the SEMAR-CONACYT project 2003-C02-11873, MEXICO.

Finally, we would like to express the honor we feel on writing this book in the year of the 150th anniversary of the publication of the dissertation of James Clerk Maxwell, *A Dynamical Theory of the Electromagnetic Field*, who has been an academic guide and inspiration for our work with antennas.

Giselle M. Galvan-Tejada
Marco A. Peyrot-Solis
Hildeberto Jardón-Aguilar
Mexico

About the Authors

Giselle M. Galvan-Tejada was born in Mexico City, Mexico. She earned a B.Sc. degree in communications and electronics engineering from the National Polytechnic Institute, Mexico, in 1994, an M.Sc. degree in electrical engineering from the Center for Research and Advanced Studies of IPN, Mexico, in 1996, and a PhD degree in electronics and telecommunications engineering from the University of Bradford, U.K., in 2000. Currently, she is working in the Communications Section of the Department of Electrical Engineering of the Center for Research and Advanced Studies of IPN, Mexico, as a lecturer and full-time researcher. She is member of the IEEE and National Researcher of the National Council of Researchers of Mexico. Her research interests include radio communication systems, wireless sensor networks, radio propagation, antenna array technology, ultra wideband antennas, WiMAX, space division multiple access, and techniques to make efficient use of the spectrum.

Marco Antonio Peyrot-Solis was born in Veracruz, Mexico. He earned a BS degree in naval sciences engineering from the Mexican Naval Academy, an MSc degree in electrical engineering from the United States Naval Postgraduate School and a PhD in communications from the Centre for Research and Advanced Studies of the National Polytechnic Institute in Mexico City in 1989, 2003 and 2009, respectively. Currently, he is working for the Mexican Navy, and his research interests are UWB antennas and electromagnetic compatibility.

Hildeberto Jardón-Aguilar was born in Tenancingo, Mexico. He earned a BS degree in electrical engineering from ESIME-IPN, and a PhD degree in radio systems from the Moscow Technical University of Telecommunications and Informatics. He has been working as a full professor at the Center for Research and Advanced Studies of IPN since 1985. His research interests include analysis of nonlinearities in RF and microwave circuits, electromagnetic compatibility, antennas and photonic systems. He is the author of five books and more than 100 technical papers published in journals and symposiums.

1

Introduction

CONTENTS

1.1   Importance of Antennas in Modern Life

Nowadays, wireless applications have become an important part of people's lives. In the field of telecommunications, for example, it is common throughout the world to find people of all ages carrying at least one wireless device with them on a daily basis. Interest in wireless technology lies in the desire for the freedom to move, while having access to information, communication and the control of different devices. Thus, wireless technology provides a flexible and attractive option for carrying out a range of tasks.

Antennas are a very important component of wireless devices, as they represent the way to receive and transmit certain signals. Basically, the antenna is the mechanism that transforms guided energy from a transmission line into radiated energy traveling through space at distances ranging from a few centimeters to hundred of kilometers. These devices have been studied and designed for a wide range of applications for over 100 years. Each application demands particular signal features, of which the required bandwidth is one of the most significant. Thus, depending on the bandwidth, three types of system can be defined: narrowband, wideband, and ultra wideband.

1.2   Ultra Wideband Systems

The need for increasingly wider bandwidths for many modern radar, imaging and telecommunications applications has propelled the search for new technologies. Several emerging lines of research sought to rise to this new challenge, and discussions took place around the world about the different ways of generating special signals to be radiated at a very large frequency range. The Ultra Wideband (UWB) concept appears to have been adopted by the United States Department of Defense in 1989 to refer to a range of terms, such as: impulse, carrier-free, and large-relative-bandwidth signals [1]. In 1992, the United States Federal Communications Commission (FCC) specified different technical standards and operation restrictions for three types of UWB systems, and marked their frequency range from 3.1 to 10.6 GHz [2]. The FCC also specifies that for an antenna (or more generally, a system) to be considered as UWB-type, it must have a bandwidth greater than 500 MHz [3].

Systems based on UWB technology transmit streams of extremely short pulses (around 10 to 10,000 picoseconds [4]), which can be spread through a very broad range of frequencies. Some interesting features of the UWB systems are obviously the possibility to carry a huge amount of information data and that this signal spread makes them robust to interference. Another important aspect is security, provided that UWB short pulses are harder to jam.

Nevertheless, at system level, the most important drawback of UWB networks is that they are range limited (between 10–20 m approximately [4]) and medium or large scale deployments cannot be implemented. Indeed, the original idea from which UWB technology was formulated was for Personal Area Networks (PANs). This implies that they can only be rolled out for short range networking, either in a centralized or distributed wireless mode (for example PANs, wireless sensor networks, etc.). For in-home purposes, for instance, UWB systems can be competitive with limited range technologies like Bluetooth and can be a complement to WiFi networks. Thus, through UWB access huge quantities of information (videos, photos, music, presentations, etc.) can be shared while avoiding cables. Naturally, UWB can also be implemented in other indoor scenarios e.g. offices, hospital rooms, academic laboratories, and industrial areas, among others.

The extremely large bandwidth of UWB systems introduces some peculiarities into the propagation and therefore in the system performance. A much more diverse multipath phenomenon is presented. Because the power is now distributed over a larger bandwidth (i.e., over many multipath components) the energy on each path could be too low to be distinguished with classical techniques. In addition, these multiple paths could suffer distinct frequency selective distortions, which could affect the pulse shapes, and as a consequence new schemes of synchronization are required. Another problem is related to the estimation of the time of arrival of UWB signals. This topic is crucial of course for applications that require high time resolution like medical imaging or radar.

Finally, the spectrum regulation is certainly another central premise of UWB systems. Basically, the frequency range assigned for these systems overlaps with other licensed and unlicensed systems, under certain spectral masks depending local regulations, which sometimes are not uniform. Generally speaking, the spectral shape of a signal is determined by the type of pulse transmitted and the modulation format [5]. Then for UWB systems, the design of pulses is also a delicate design subject. In this concern, the chapter *Ultra wideband pulse sharper design* can be consulted [6].

1.3 UWB Antennas

The study of UWB antennas is important because, as mentioned in the previous section, UWB systems transmit and receive ultra-short electromagnetic pulses, which means that they use a very wide bandwidth with low levels of average power, creating difficulties with signal detection; to overcome this, UWB antennas need to receive all the components of the signal spectrum with the same efficiency, and without introducing significant distortion in the phase of those frequency components. Although it is possible to implement a type of compensation as a countermeasure to distortion effects [7], it is far preferable to seek antenna designs that introduce the lowest possible distortion in the phase characteristic, such that the whole system design is simpler.

Therefore, the behavior and performance of an ultra wideband antenna must be coherent and predictable throughout the operational bandwidth, which means: variations must not be introduced into the antenna radiation pattern (failing this, there should be as few variations as possible), there must be good matching conditions (evaluated by the reflection coefficient), and there must be no distortion of the signal waveform. Moreover, besides traditional parameters to describe narrowband antennas (e.g., gain, impedance matching, polarization, etc.), other parameters must be considered, such as phase linearity and radiation pattern variations across the frequency range, which are important for the satisfactory integration of antennas into modern UWB systems.

UWB antenna theory presents an additional challenge compared to classical antenna theory. In the latter, theoretical developments are based on knowledge of the wavelength, as determined by its main resonance frequency; however, the main resonance frequency of UWB antennas is non-determinable, since more than one resonance can be identified in their very broad bandwidths. This characteristic leads, therefore, to uncertainty as to the use of the lower cut-off, the upper cut-off, or any "central" frequency.

In the open literature today there are more than 6000 articles related to UWB antennas, a fact that demonstrates the relevance of the topic. In particular, significant efforts have been focused on the development of omnidirectional antennas, due to their implementation in mobile applications. However, UWB antennas with directional radiation features have also attracted the attention of researchers and manufacturers, since these antennas are particularly attractive for a military environment.

1.4   Scope of the Book

The present text was designed to be reference material for the interested reader, and it is important here to outline the scope of the book. In Chapter 2, general concepts relating to classical antenna theory and design are presented, with a brief explanation of the main parameters of "conventional" antennas that have been used for many years.

Chapter 3 is a compendium of several recent developments in ultra wideband antennas reported in different forums. These developments are grouped according to the identification of their structures. In general terms, two types of structures are defined: planar and planarized. (The 'planarized' term is not usually used in the literature, but we introduce it in order to differentiate between ground plane embedded and non-embedded flat radiators). Some early work that intended to achieve a wider operational frequency range is also covered in this chapter, providing many guidelines on the fundamentals of UWB antennas.

Theory developed for UWB antennas is addressed in Chapter 4, where the first parameter to define is bandwidth. As pointed out in this chapter, although it is possible to directly formulate a single definition for bandwidth, there are many founding concepts behind this. A fundamental concept for ultra wideband antennas is the quality factor, which was studied by Wheeler in the context of small antenna [8,9] and was also explored by Schantz, who analyzed the dependence of this factor on bandwidth [10]. This factor has an inverse relation with bandwidth, which means that UWB antennas must have low quality factors in order to store the least possible energy, thus achieving wide bandwidths. Another important concept is solid-planar equivalence, which allows planar and planarized structures to be implemented instead of volumetric antennas. This principle is vital for the development of UWB antennas in portable devices, and for the recent body area networks where small, low profile antennas are intended to be integrated in wearable devices.

A particular aspect of UWB antennas relates to time domain signal analysis. Due to the relative short duration of UWB pulses, the transient response must not be neglected since it provides a measure of the dispersion (and in general, the distortion phenomenon) that they suffer. The importance of having a measure of this phenomenon is that it imposes certain constraints on the data rate to be transmitted. These issues, together with performance measurements such as group delay and phase linearity, are addressed in Chapter 5. Examples of dispersive and non-dispersive antennas are also given in this chapter.

Chapter 6 presents design guides for UWB omnidirectional antennas, where the challenge is to conserve the radiation pattern shape over the whole bandwidth. A design methodology is described, which considers three objectives: impedance matching, evaluated through the reflection coefficient magnitude; phase linearity, determined through the phase of the reflection coefficient; and variations in the shape of the radiation pattern. In essence, four parameters related to radiator dimensions are the variables within the methodology. The chapter includes different omnidirectional UWB designs as reported in the open literature, and a performance comparison among them.

Directional UWB antennas are covered in Chapter 7. This chapter also presents a specific design methodology, which is based on the methodology of omnidirectional antennas explained in Chapter 6. The difference here is a new variable for the inclination angle of the radiator. This radiator tilt is precisely the mechanism that allows directive radiation patterns to be achieved. A different design methodology based on the solid-planar correspondence principle is also addressed in this chapter, which allows to generate a structure that can be designed for any lower cut-off frequency. Simulation results of UWB directive antennas reported by some authors are compared, in order to evaluate their performance.

It is worth noting that all simulations presented in this book have been conducted using the CST Microwave Studio. The measurement results shown in Chapters 6 and 7 were carried out with an Agilent network analyzer model E8362B.

Some current trends and unresolved problems in the field of UWB antennas are addressed in Chapter 8. Of particular interest are the relatively recent body area networks, where a vast variety of antennas have been designed. Medial imaging is another topic included in this chapter, oriented mainly to breast cancer detection, where a high temporal resolution must be achieved. The principal features of some antennas proposed for this aim are presented.

The book concludes with Chapter 9, where an important subject related to analysis and design of UWB antennas is covered: Numerical Methods for Electromagnetics. The complexity of Maxwell's equations and how to solve them has been a matter of interest for diverse authors for many years. Numerical tools with different approaches have been developed in this field. The generalities, origin, and steps of three popular methods, Finite Differences, Finite Element Method and Method of Moments are discussed. Some commercial software for simulating antennas and their capabilities are also included.

Bibliography

[1] T. W. Barret. History of ultra wideband (UWB) radar & communications: pioneers and innovators. In *Progress in Electromagnetics Symposium 2000 (PIERS2000)*, 2000.

[2] FCC. First report and order, revision of part 15 of the commission's rules regarding ultra-wideband transmission systems. Technical report, Federal Communications Commission, 2002.

[3] FCC. US 47 CFR part 15 subpart F §15.503d ultra-wideband operation. Technical report, Federal Communications Commission, 2003.

[4] E. K. I. Hamad and A. H. Radwan. Compact UWB antenna for wireless personal area networks. In *2013 Saudi International Electronics, Communications and Photonics Conference*, pages 1–4, 2013.

[5] J. G. Proakis. *Digital Communications*. Mc-Graw Hill, Boston, MA, 3rd edition, 1995.

[6] Z. Tian, T. N. Davidson, X. Luo, X. Wu, and G. B. Giannakis. *Ultra Wideband Wireless Communication*, Chapter 5, pages 103–130. John Wiley & Sons, 2006.

[7] T. W. Hertel and G. S. Smith. On the dispersive properties of the conical spiral antenna and its use for pulsed radiation. *IEEE Transactions on Antennas and Propagation*, 51(7):1426–1433, 2003.

[8] H. A. Wheeler. Fundamental limitations of small antennas. *Proceedings of the IRE*, 35(12):1479–1484, 1947.

[9] H. A. Wheeler. The radiansphere around a small antenna. *Proceedings of the IRE*, 47(8):1325–1331, 1959.

[10] H. Schantz. *The Art and Science of Ultra Wideband Antennas*. Artech House, Norwood, MA, 2005.

2

General Concepts of Antennas

CONTENTS

2.1 Introduction

In this chapter some general concepts associated with antennas are introduced. These concepts are valid for both narrowband and non-narrowband antennas. It is not the objective of this book, however, to present a detailed theory in the field of antennas. A wide variety of excellent references for theory and practice on classical narrowband and broadband antennas can be found in the open literature [1–5]. The particularities of the case of UWB antennas will be addressed in Chapter 4. In order to provide some fundamentals on these devices, the most representative classical narrowband antennas are briefly presented below.

2.2   Classical Narrowband Antennas

2.2.1   Wire antennas

One of the basic elements needed to produce electromagnetic radiation is a single wire through which a certain electrical current flows [1]. In order to achieve radiation of a particular strength, different single or complex structures for the antennas have been designed. In the following subsections some elementary radiators are briefly described.

2.2.1.1   Dipole

When the conductors of a transmission line are opened 90° relative to its axis and each conductor is in an antipodal direction, a *dipole radiator* is formed as shown in Figure 2.1. The phenomenon of radiation achieved by this change in a transmission line, as well as its associated theory, are detailed in [1]. In practice, dipole antennas are designed as a 'sole' device, i.e., independent of the transmission line that feeds it.

2.2.1.2   Hertzian monopole

By considering only one branch of the dipole antenna, it is possible to have a single wire radiator known as a *Hertzian monopole*. This antenna achieves the behavior of a dipole antenna provided that a mirror image is formed with the radiator element by means of a ground plane. The geometry of this type of antenna is shown in Figure 2.2. Much theory has been developed around this basic element (e.g., [1]) and its simplicity and electrical attributes (see Section 2.5 below) have allowed it to be used for many applications around the world.

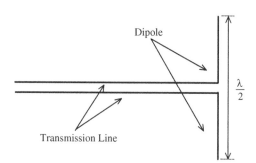

FIGURE 2.1
Dipole antenna.

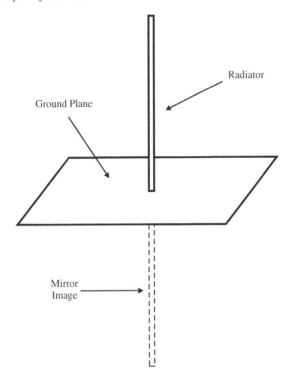

FIGURE 2.2
Monopole antenna.

2.2.1.3   Loop antennas

It is also possible to fold a wire in such a way that a loop is made, as shown in Figure 2.3, where two possible configurations of a loop antenna are depicted. An antenna known as a *folded dipole* has been used as a radiator element in Yagi-Uda antennas used in television broadcasting. This type of antenna is also useful for magnetic sensors.

2.2.2   Aperture antennas

The aperture antennas operate based on the principle that an electromagnetic field is distributed over the surface of a cavity. This is the typical situation of the electromagnetic field propagating in a waveguide. In the case of antennas, let us say that the guide is sawn off in such a way (in a plane perpendicular to the guide's axis) that it ends in a particular "mouth," hence the term aperture. Just at this edge is where the radiation is produced. Now, due to the fact that energy is concentrated in this structure (i.e., the sawn off guide, better known as *the horn*), the antenna radiates energy concentrated in a particular direction. This type of radiator is used mainly at microwave frequencies and

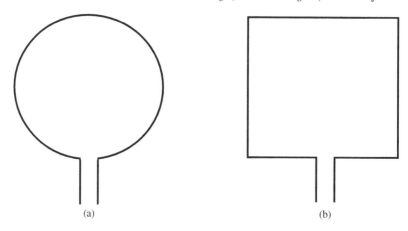

FIGURE 2.3
Two loop antenna configurations: (a) Circular loop, (b) rectangular loop.

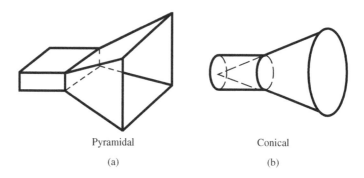

FIGURE 2.4
Antenna horn types: (a) Pyramidal, (b) conical.

the basic structures are the pyramidal and the conical horns as shown in Figure 2.4.

2.2.3   Reflectors

A possible way to concentrate the energy radiated (or received) is through reflectors and lenses. Let us assume a parabolic reflector such as depicted in Figure 2.5. If a basic radiator element, such as a horn or dipole (usually known as a *feeder*), is placed at the focal point of the parabola, it will radiate the energy oriented according to the parabolic axis and concentrated in the reflector aperture. There are several types of reflectors and lenses used for this purpose, e.g., cylinder, corner, and cut paraboloid, among others, with the full paraboloid being the most popular.

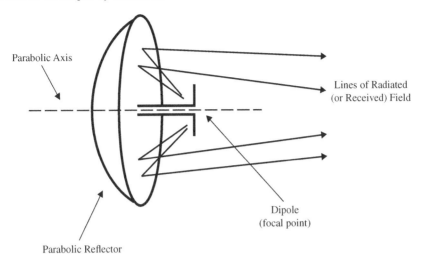

FIGURE 2.5
Parabolic reflector.

2.2.4 Microstrip antennas

Microstrip antennas emerged from the miniaturization process in wireless electronic devices. They are basically a metallic flat patch printed over a dielectric surface (with a certain associated relative permittivity, ε_r). Figure 2.6 shows an example of a microstrip antenna (sometimes called *patch antenna*) with a square shape. Although different shapes are possible (triangular, elliptical, rectangular, circular ring, etc.), the most popular shapes are the square and the circular patches due to their relatively simple design and performance. As well as the patch and the dielectric, these antennas need a ground plane, which is a thin metallic surface located parallel to the patch and under the dielectric, as illustrated in Figure 2.6.

These antennas have become very popular in space-limited applications such as aircraft, mobile devices, cars, and terminals of body area networks, among others, and for frequencies of microwaves and higher. However, they are, in principle, narrowband devices (see Section 2.8, below) and hence have certain constraints for use in broadband applications.

2.2.5 Antenna arrays

The basic objective of forming an antenna array is to concentrate the received or transmitted energy in a particular direction, providing a higher gain and directivity than that achieved with a single element (see Section 2.4 below on concepts of gain and directivity). Thus, an antenna array is a structure made up of a certain number of elements and arranged in a specific configuration.

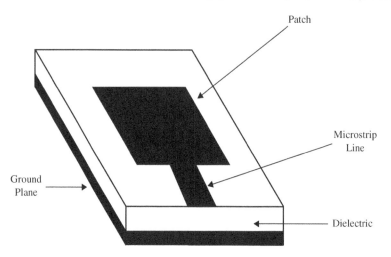

FIGURE 2.6
Microstrip antenna.

It is worth noting that the elements can be any of the antennas presented in the previous subsections 2.2.1, 2.2.2, and 2.2.4.

The radiation features of an array obey the fact that the individual responses of each element are combined in such a way that it is possible to fulfill the concentrated energy effect (principle of *pattern multiplication* [3]). Hence, the response of the array, in terms of its radiated energy, depends on the distance between adjacent elements, d, the magnitude and the phase of the currents on each element, and the individual radiation property of each element. Figure 2.7 shows a representation of an antenna array where there are in general N elements along a line. In any type of array configuration one element is always considered (in the case of the array of Figure 2.7, the element 1) in such a way that at a particular observation point, let us say P, it is possible to determine the field from each element and its contribution to the total field.

There are two possible ways in which an array can radiate: one, when maximum radiation occurs in a direction perpendicular to the array line, in which case it is said that the antenna is of the *broadside* type; or two, when the maximum radiation is concentrated in the direction of the array line, there is an *end-fire* array. Their nature depends on d and the relative difference of the phases, α, in each element of the array. The mathematical expressions to obtain the response of an array as a function of these variables as well as the number of elements is beyond of the scope of this book, but there are several classical references for the interested reader, [1] and [6].

Although up to now the explanation of the array has been based on a single line structure, there are other geometries such as circular, triangular,

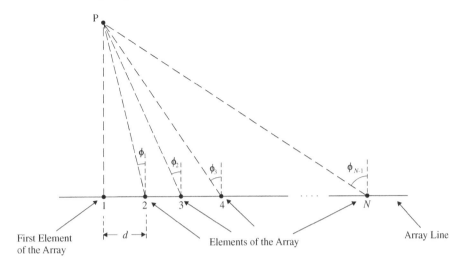

FIGURE 2.7
Antenna array representation.

square, or a matrix of elements arranged in a particular configuration. It is also possible that there are distinct distances between elements, although it is not a common practice to implement such topology. When the elements are arranged on a single line and they are uniformly distributed, this array is known as *uniform linear array* (ULA), and a wide variety of articles have been published in the open literature presenting different studies and results based on a ULA antenna. Another very popular array is the Yagi-Uda antenna used for broadcasting and which is briefly explained below.

2.2.5.1 Yagi–Uda

The Yagi–Uda antenna (name given by its creator Uda whose Japanese work was described in English by Yagi) is an end-fire type array, but instead of feeding each of the elements, it exploits the fact that the energized or driven element induces a field to adjacent elements by means of a *parasitic excitation process* (hence these elements are known as *parasitics*). Now, depending on the space between the parasitics elements (usually five or six [7]), these will produce an increase in the radiated field of the driven element, so concentrating the energy in an opposite direction to this element. For this reason the parasitic elements are also known as *directors*.

In addition, the Yagi–Uda antenna has another element located on the other side of the directors and adjacent to the driven element (see Figure 2.8). Depending on the design of the element itself (both size and position relative to the driven element), it will reinforce the radiated energy toward the directors. Thus, this element is known as a *reflector*.

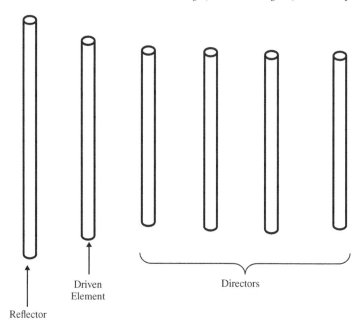

FIGURE 2.8
Yagi–Uda antenna.

2.3   Feed-Point

In all cases of antennas presented in Section 2.2 and any other type of antenna, these devices need to be connected to a radiofrequency generator (transmitter) or to a receiver. The interface between both elements can be a transmission line or a waveguide. In any case, there is a place where the transmission line is physically connected to the antenna, which is known as the *feed-point* (see Figure 2.9) and depending on its design, different antenna responses can be obtained.

2.4   Gain and Directivity

One of the most important concepts associated with antennas is their gain, which will provide a general idea of the type of antenna used by designers and operators of wireless systems. In a general context, the term gain is usually explained based on a comparison of two quantities. In the case of antennas,

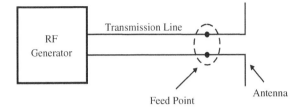

FIGURE 2.9
Feed-point.

a reference antenna is used to compare its performance with a corresponding antenna whose gain will be determined (let us call it *antenna under study*). So before continuing with the concept of antenna gain, let us address the term isotropic antenna. An isotropic antenna or isotrope is an ideal radiator, which is non-physically realizable. The energy "emitted" by this antenna is uniformly distributed over an imaginary sphere of radius r, in other words, the field strength is the same at any point of the sphere surface, and consequently an isotrope is used as a reference antenna. Figure 2.10 depicts the geometry of the radiation sphere where it is assumed that the isotropic antenna is at the origin of the spherical coordinate system. As can be seen in this figure, the radiated energy at an arbitrary point P is a function of two angles, φ and θ.

When the power generated by any other antenna is compared to that of the corresponding isotropic radiator (i.e., the difference of radiated power produced by both antennas is obtained), the gain produced by the antenna under study is determined. Provided that the isotropic antenna is used as a reference, it is said to have a gain equal to one in a linear scale or 0 dB in a logarithmic scale. In general terms and mathematically, the antenna gain is given by

$$G(\varphi, \theta) = \frac{P_t(\varphi, \theta)}{P_i(\varphi, \theta)} \tag{2.1}$$

where $P_t(\varphi, \theta)$ is the power radiated (or received) by the antenna under study and $P_i(\varphi, \theta)$ the power "radiated" (or received) by the isotropic radiator, both for a certain pair of angles φ and θ. Please note that in the spherical coordinate system one should include the variable of distance r. However, provided that the antenna gain is obtained for the same value of r in all possible cases of φ and θ, Equation (2.1) is valid.

As regards the directivity, this term is sometimes confused with the gain concept explained above. This confusion usually occurs because specifications of commercial antenna data sheets provide, among other parameters, a certain value of gain for the antenna to be consulted, but which corresponds to the maximum gain achieved by an antenna. This value is in fact the directivity of an antenna.

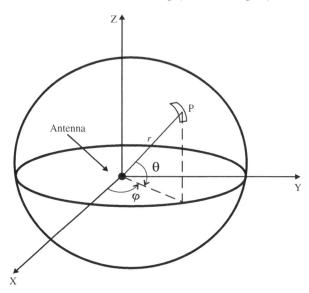

FIGURE 2.10
Radiation sphere of an isotropic antenna.

2.5   Radiation Pattern Concept

The radiation pattern is a graphical representation of the energy distribution around the antenna at a particular distance. In principle, any antenna radiates energy in a three dimensional space. The simplest situation is given by an isotropic antenna, whose resulting radiation pattern is a sphere, as introduced in Section 2.4. In practice, it is common that commercial antenna specifications show the radiation pattern only in the horizontal or azimuthal plane and the vertical plane (i.e., when $\theta = \frac{\pi}{2}$ and $\varphi = 0$, respectively). For example, Figure 2.11 shows the three dimensional radiation pattern of a $\lambda/4$ monopole obtained by simulations using the CST Microwave Studio simulator.

So far, the concepts of gain and radiation pattern have been explained in a radiation context only (indeed, the energy distribution of an antenna is summarized under the term radiation pattern). Nevertheless, due to the reciprocity principle of antennas, an antenna is able to receive energy with the same gain and efficiency in a particular direction as it radiates. Therefore, these concepts are applied equally if an antenna is radiating or receiving energy. This is also applied to the parameters presented in the following sections.

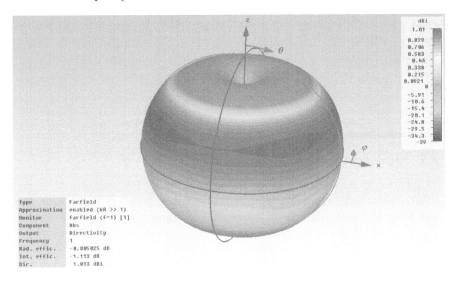

FIGURE 2.11
Simulated three-dimensional radiation pattern for a monopole.

2.6 Polarization

As is well known, a radiated electromagnetic wave is made up of an electric component (**E** vector) and a magnetic component (**H** vector), which are perpendicular each other. The resulting vector, perpendicular to the plane formed by **E** and **H** (*Poynting vector*) indicates the propagation direction of the electromagnetic wave.

Now, the polarization of a wave corresponds to the direction of the **E** vector, so that if this vector lies in a vertical plane, the wave is vertically polarized, whereas horizontal polarization results if the wave is horizontal or lies in a horizontal plane. Each of these situations depends on the orientation of the radiating element of an antenna (vertical or horizontal).

In any case, when the **E** vector has a particular polarization for all values of the propagation direction of the electromagnetic wave, it is said that it has a *linear polarization* (and therefore it is understood that the signal comes from an antenna also linearly polarized).

There is another type of polarization that occurs when two orthogonal linear polarized waves are combined and radiated simultaneously by the same antenna. If the magnitudes of the two components of **E** are equal, there is a circular polarization, otherwise the polarization is *elliptical*.

The polarization is a parameter, which indicates how the antenna must be designed in order to match the signals of the transmitter-receiver pair. In other words, if a transmitting antenna has vertical polarization, the antenna that is receiving its signal must also have vertical polarization. In fact, by exploiting this property (i.e., different types of polarizations), it is possible to make efficient use of the spectrum and provide a diversity mechanism.

2.7   Impedance

As mentioned in Section 2.3, any antenna is fed by its generator through a transmission line, which is connected at the feed-point of the antenna. At this point the antenna presents a certain impedance to the transmission line. Therefore, it is known as *load impedance* Z_L (as is seen by the transmission line) and it also represents the antenna input impedance Z_i. If $Z_L = Z_0$, where Z_0 is the characteristic impedance of the transmission line, there will be maximum energy transference and non standing wave will be present on the line. Otherwise, there is an impedance matching problem.

We now denote voltage and current at the feed-point (i.e., at the antenna input), as V_i and I_i, respectively. Then, the input impedance is given by,

$$Z_i = \frac{V_i}{I_i} \qquad (2.2)$$

As is well known, Z_i is, in general terms, a complex quantity made up by resistive and reactive components. If the reactive component is dominant and has a large value, the voltage applied to the antenna input (V_i) must also be large such that the antenna can produce enough radiated power. Thus, the larger the value of Z_i, the larger the level of V_i such that the antenna allows a larger current I_i to flow. Consequently, it is desirable to have as low a Z_i value as possible.

2.8   Reflection Coefficient

As has been addressed previously, it is essential that the antenna have an adequate input impedance, so that standing wave is avoided when the antenna is connected to the transmission line. Then, depending on the quality of the impedance matching between the antenna and the transmission line (i.e., at the feed-point), the energy traveling from the transmitter to the antenna could be reflected causing a reduction in the radiated energy of the antenna. The ratio of the amplitude of the reflected wave, V_r, to the amplitude of the incident wave, V_i, at the feed-point is known as *reflection coefficient*, Γ:[1]

$$\Gamma = \frac{V_r}{V_i} \qquad (2.3)$$

According to Equation (2.3), Γ can take a maximum value of one, which would occur if $V_r = V_i$ and a total reflection is presented. Consequently, it is desirable that Γ be as small as possible.

Let us now address some aspects of the impedance matching between the antenna and the transmission line. From the relationship of the antenna load impedance, Z_L, and the characteristic impedance of the line, Z_0, the value of Γ can be entirely determined. This relation is derived in [3] and is given by

$$\Gamma = \frac{Z_L - Z_0}{Z_L + Z_0} \qquad (2.4)$$

From Equation (2.4) it is seen than when $Z_L = Z_0$, $\Gamma = 0$ and therefore there is a total energy transference to the antenna terminals, as was shown in Section 2.7. Provided Equation (2.4) is an impedance relationship, it is clear that Γ must be a complex quantity. Thus, Γ can be represented by

$$\Gamma = |\Gamma| e^{(j\phi)} \qquad (2.5)$$

where $|\Gamma|$ and ϕ are the magnitude and the phase angle of the reflection coefficient, respectively. Then, the physical meaning of Equation (2.3) corresponds to $|\Gamma|$, i.e., strictly speaking, Equation (2.3) must be rewritten as

$$|\Gamma| = \frac{V_r}{V_i} \qquad (2.6)$$

and care should be taken to use Equation (2.6) instead of Equation (2.3).

As regards the phase angle, it represents the difference between the phases of the incident and reflected waves, i.e.,

$$\phi = \phi_i - \phi_r \qquad (2.7)$$

When the load is not perfectly matched to the transmission line, the reflections at the load cause the traveling wave to propagate in an inverse direction to the incident wave. In this way, a standing wave pattern is created in the transmission line, whose impact can be characterized through a ratio of the maximum and minimum amplitudes of the traveling waves, V_{max} and V_{min}, respectively, known as *VSWR* (*Voltage Standing Wave Ratio*) and can be determined by [3]

$$VSWR = \frac{V_{max}}{V_{min}} = \frac{1 + |\Gamma|}{1 - |\Gamma|} \qquad (2.8)$$

The maximum value of $VSWR$ that indicates a good impedance matching is 2, therefore any $VSWR$ value less than 2 is highly desirable.

---

[1]It is worth mentioning that this coefficient is also denoted as S_{11} from the scattering parameters or S-parameters theory.

2.9   Quality Factor

In the context of circuit theory, a resonant RLC circuit (resistor, inductor, capacitor) presents what is known as *Quality Factor*, or simply *Q factor*, which represents the proportion of stored energy to the dissipate energy of the circuit. Consequently it is desirable that a circuit leaks the least energy possible, which is achieved through increasing its reactive part. Nevertheless, the physical operation of an antenna (which can be described as a resonant RLC circuit) dictates that this device must be able to radiate or dissipate its energy. This means that the higher the Q factor of an antenna, the more reactive its input impedance becomes, and therefore the radiated energy is concentrated in a narrower bandwidth. Meanwhile the lower the Q factor, the wider the bandwidth due to the increase in the radiated energy distributed in the frequency span.

The expression to determine Q is known as the fundamental performance limit of an antenna provided that Q limits the antenna bandwidth [8], and is given by [9]

$$Q = \frac{1 + 2(kR)^2}{(kR)^3 \left[1 + (kR)^2\right]} \tag{2.9}$$

where R is the radius of an imaginary sphere that contains the antenna and k is the wave number ($k = 2\pi/\lambda_c$ with λ_c the wavelength of the central frequency).

2.10   Bandwidth

The bandwidth of an antenna corresponds to the frequency interval within which the antenna operates satisfactorily. In this sense, the first idea related to this parameter is that there must be a maximum and a minimum frequency that limit the bandwidth, which is indeed the case. The difference between both frequencies (usually called *cut-off frequencies*) provides the bandwidth.

The most controversial aspect of this is the meaning of the *satisfactory operation* of the antenna, and hence it is necessary to have a parameter which is a function of the frequency and for which the lower and upper cut-off frequencies can be defined.

There are distinct antenna parameters whose performance is a function of the frequency. As was explained in Section 2.8, the reflection coefficient provides a measure of the antenna matching. Thus, this parameter is fundamental to determine antenna bandwidth and often the specifications sheets of commercial antennas show the bandwidth of the device, which is obtained from the magnitude of the reflection coefficient. It is worth noting that when the

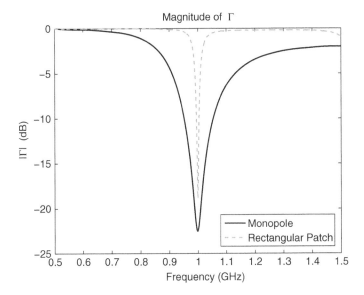

FIGURE 2.12
Simulated magnitude of Γ for a wire monopole and a rectangular patch antenna.

bandwidth is related to Γ, it is known as *impedance bandwidth*. Some antennas parameters are associated with the impedance bandwidth: input impedance, radiation resistance and efficiency [3].

The bandwidth of an antenna can be narrowband, broadband or the recent ultra wideband such as those to be presented in Chapter 3. In addition, depending on the narrowness or broadness of the bandwidth, it is specified as a percentage of a central frequency (i.e., the operational frequency) or as a ratio of the maximum to the minimum frequency, respectively.

When that ratio is 2:1 or less, the bandwidth can be given as a percentage of a central frequency and the antenna is narrowband, otherwise, the antenna is of the broadband type because a percentage is not possible (nor a central frequency) and it is specified by its ratio of frequencies, as will be explained in Chapter 4.

In order to obtain the bandwidth of an antenna it is common that a threshold of -10 dB is established in the magnitude of the reflection coefficient (or equivalently a $VSWR < 2$). Figure 2.12 shows the simulated magnitude of Γ of a wire monopole and a microstrip antenna operating in the 1 GHz band, for illustration purposes. By considering a threshold of $|\Gamma| = -10\,\text{dB}$, it can be seen that the microstrip antenna presents a bandwidth narrower than that of the wire monopole.

The behavior of the radiation pattern through a certain frequency range also dictates the antenna bandwidth; specifically, the interval of frequencies where the radiation pattern keeps its shape or is almost unchanged. Then, when the bandwidth is evaluated as a function of the radiation pattern response, it is known as *pattern bandwidth*. The parameters associated to this bandwidth are gain, beamwidth, side lobe level and polarization [3].

Bibliography

[1] C. A. Balanis. *Antenna Theory: Analysis and Design.* John Wiley & Sons, 3rd edition, 2005.

[2] W. L. Stutzman and G. A. Thiele. *Antenna Theory and Design.* John Wiley & Sons, 1998.

[3] L. V. Blake. *Antennas.* Artech House, 1984.

[4] R. S. Elliot. *Antenna Theory and Design.* Prentice-Hall, New Jersey, 1981.

[5] R. E. Collin and F. J. Zucker. *Antenna Theory.* McGraw-Hill, New York, 1969.

[6] M. T. Ma. *Theory and Application of Antenna Arrays.* John Wiley & Sons, 1974.

[7] American Radio Relay League. *The ARRL Antenna Handbook*, 1991.

[8] R. C. Hansen. Fundamental limitations in antennas. *Proceedings of the IEEE*, 69(2):170–182, 1981.

[9] H. Schantz. *The Art and Science of Ultra Wideband Antennas.* Artech House, Norwood, MA, 2005.

3

Recent Developments in Ultra Wideband Antennas

CONTENTS

3.1   Introduction

The impact of wireless communications in recent years has been remarkable thanks to technological developments which have allowed more and more users to carry one or more wireless device to carry out a range of activities through radio communication systems. In these systems, the antenna is certainly a key element.

As pointed out in Chapter 2, an antenna can be described, in a broad sense, as a transducer that converts the electromagnetic energy guided into a transmission line into radiated electromagnetic energy [1]. Most antennas are reciprocal devices, and behave in the same way both to transmit and to receive; however, the antennas are treated as either transmitting or receiving devices according to each status. In reception mode, for instance, the antennas act as collectors of the arriving electromagnetic waves, which are directed toward a common point for their processing. In some cases antennas focus radio waves in the same way as eyeglasses focus optical waves [2].

These general definitions are applicable to both narrowband and broadband antennas. However, a more specific definition for antennas of ultra wideband (UWB) establishes that these antennas are non-resonant radiators of a low quality factor Q (which is a factor associated with resonant structures, see Chapter 2), and whose input impedance remains quasi-constant on a very wide operation frequency band [3]. Naturally, and like other types of antennas, the UWB antennas need to radiate as much energy as possible, avoiding waves reflected toward the transmission line in their operational bandwidth.

The end of the Cold War allowed technologies that had previously only been used in the military environment to be transferred to commercial applications (e.g., communication systems of spread spectrum by code division like direct sequence and UWB systems). In the year 2002, the Federal Communications Commission (FCC) of the United States specified different technical standards and operation restrictions for three types of UWB systems (radars for vehicles, communication systems, and measurement and imaging systems) and also stipulated that the frequency span for the UWB systems be from 3.1 to 10.6 GHz [4].

This wide bandwidth imposes certain operation conditions on the UWB systems. Basically, these systems transmit and receive ultra short electromagnetic pulses. The above means they use signals with an ultrawide bandwidth with very small transmission power. Among other outcomes, this property makes signal detection and interception more difficult, which is why UWB technology has been widely used in military applications since the 1960s (as introduced in Chapter 1, other names were given to this technology). Hence, a UWB system requires the antenna to be able to receive all the signal spectrum components efficiently, and without introducing a significant distortion in the phase of these components. Therefore, the behavior of the antenna and its performance should be consistent and predictable throughout the whole operation band. This means that ideally, the radiation patterns and their matching characteristics should be stable through the whole band. Also, a UWB antenna should preferably not introduce pulse distortion, although if the waveform dispersion is presented in a predictable way, it should be compensated. [1]

The development of UWB technology has shown that the traditional parameters to describe an antenna, such as gain, impedance matching, polarization, etc., are specified in a precise way only for narrowband antennas, since to describe the UWB antennas needed for this technology, additional parameters are needed such as phase linearity, stability of the radiation pattern, etc., which are important factors for their appropriate integration into modern communication systems [5].

One of the most important features in the UWB antennas addressed in this book is directivity. As is well known, compared to an omnidirectional antenna, a directional antenna concentrates energy into a narrow solid angle. In other words, the density of electromagnetic energy is radiated from the antenna with an intensity that varies according to the angle around the antenna [2]. Generally, a directional antenna needs to be of a relatively larger size compared to an omnidirectional antenna.

Based on the previous paragraph, it may seem that an initial classification of UWB antennas could be between directional and omnidirectional antennas, which would provide an initial approach to identifying their current situation. However, due to the fact that most current research has been focused on UWB omnidirectional antennas (because all efforts have fundamentally been directed toward the improvement of the wireless communications industry, centering on mobile devices that require omnidirectional radiation patterns), a better classification to present the current status of these antennas would be based on the structure of the antenna, which can be volumetric or non-volumetric.

As such, before the 1990s, almost all proposed UWB antennas were of volumetric structure. From 1992 onwards, non-volumetric UWB antennas have been proposed (e.g., printed circuit, slot and planar monopoles). So, based on this classification and according to the historical progress of UWB antennas, the volumetric antennas are first to be addressed here, whose developments provide the background for current UWB antennas. Then, a wide variety of non-volumetric antennas proposed by different authors are presented. The double ridged guide horn antenna deserves special attention, as it represents a standard UWB antenna which is useful for characterizing the radiation patterns of new designs. Therefore it is addressed in a separate section.

---

[1] A periodic logarithmic antenna is a good example of a "dispersive" antenna as will be explained in Chapter 5. [1]

3.2   Background to Ultra Wideband Antennas

The concept of UWB originates from spark-gap transmitters, forerunners in radio transmission technology. However, the designs of UWB antennas at the end of the nineteenth century were not appropriately preserved, and many of them were forgotten and later rediscovered in the middle of the twentieth century.

The first references to UWB antenna designs date back to 1898, when the North American patent 609,154 granted to the English citizen Oliver Lodge makes reference to some capacitive areas to be used in the telegraph of the Hertzian wave [6]. Although in that patent the areas are shown both in square and circular shape, a triangular shape is recommended. Lodge's design presented in this patent is the predecessor of the bow tie antenna, similar to the bi-conical discussed later.

Unfortunately, as the operational frequencies of communication systems were increased, Lodge's design was replaced by $\lambda/4$ monopoles, which provided a better performance. However, the arrival of television generated the need to investigate designs of antennas that could operate over wider bandwidths, providing a new support for the broadband antennas, which finally culminated in the rediscovery of the bi-conical antenna by P. S. Carter, who obtained the North American patent 2,175,252 in 1939 [7]. In that patent, Carter mentions that this antenna is of a short wave type, which shows neglected reactance in an extremely wide frequency band, and that therefore it can be used for television signal transmission [8].

Besides the rediscovery of the bi-conical antenna, P. S. Carter was the forerunner in the use of broadband transition between a feeder and the radiant elements, by proposing that the transmission line gradually increased its dimensions until this joined to the cone, which was in fact an enhancement of Lodge's original design, as can be seen in Figure 3.1 [7, 9].

In 1940, Sergei A. Schelkunoff proposed an antenna made up of conical wave guides and power supply structures in conjunction with a spherical dipole. These structures were of a relatively small size and presented low resistance to the air, which made this type of antenna interesting for airships use [10]. However, they were not very successful in the end, probably due to the complexity of their construction (see Figure 1.13 of [7]).

During that time there was a tendency to construct complex UWB antennas. In 1941, Lindenbland proposed an antenna which consisted of an element of coaxial horn horizontally polarized. Lindenbland argued that this antenna was optimal for transmission of a high quality television signal, for which it was finally selected by the Radio Corporation of America (RCA) to experiment with television transmissions. RCA selected this antenna as they saw the possibility of transmitting several television channels from a central station, which would require a broadband antenna with these features [7, 11].

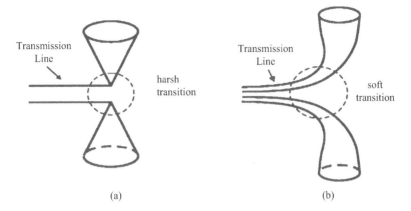

FIGURE 3.1
Transmission line transitions for (a) Lodge's design and (b) Cartes's design.

Based on this idea, other researchers built several antennas based on the coaxial transition. L. N. Brillouin, for example, patented two coaxial horn antennas with vertical polarization in 1948, one directional and another omnidirectional, which showed a terminal impedance similar to that of the free space. The directional antenna maintained a constant gain in a wide band of frequencies [7, 12].

During this decade, other researchers also developed different horn designs. For example, M. Katzin, who obtained a patent in the United States in 1946 for an electromagnetic horn radiator that increased gain, as well as obtaining large effective apertures and compensating for distortions by using horn arrays. In this context, in 1942, A. P. King patented a conical horn antenna, whose efficiency was increased while its directivity was conserved [7, 13]. Although previously existing designs had excellent performance, other factors began to gain importance. For example, when broadband receivers became common, it was considered necessary that the new antenna designs be cheap and easy to build. As a result of these new requirements, the bow tie antenna originally proposed by Lodge in 1898 was again studied in 1952 by G. H. Brown and O. M. Woodward [7]. Another antenna designed in the 1940s and that had recently been the subject of analysis, was that patented by R. W. Masters in 1947, which is simply a diamond shaped dipole with a reflector. Its main features are that it maintains a constant input impedance over a very wide bandwidth, and that due to its simple construction and its heavy-duty use, it can be easily designed for a particular application. In Figure 3.2 a drawing in perspective of this antenna can be seen [7, 14].

By the 1960s, more sophisticated antennas were being developed such as those proposed by W. Stohr, which are formed by monopoles and circular or ellipsoidal dipoles [7, 15]. Also during this decade, in 1962 G. Robert-Pierre Marié [16] developed an ultra broadband decade slot antenna, which was

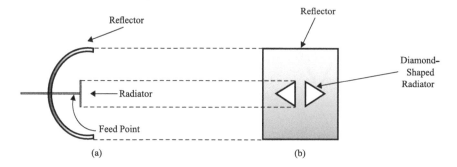

FIGURE 3.2
Representation of the design of a diamond-shaped dipole antenna proposed by R. W. Masters in 1947. (a) Lateral view, (b) frontal view.

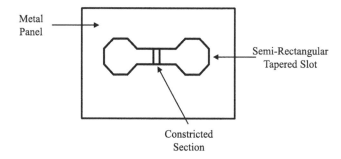

FIGURE 3.3
Broadband slot antenna. Modified from [16].

achieved by varying the width of the slot, in such a way that a small section of the radiator emits at high frequencies, whereas a large scale portion of it radiates at low frequencies. Thus, the *tapered-feed* idea was applied to achieve wider bandwidths of antennas (see Figure 3.3). Some variations of this single slot antenna were also explored by Marié. For example, one of his variations consisted of using two slots implemented on the same conductive panel, and where there is an opening at their center in order to achieve certain response at low frequencies. It is worth noting here that, as will be addressed in Chapter 5, these antennas can introduce dispersion due to the strong dependence of their phase center with the frequency [17].

During the 1980s, other antennas were under development, mainly easy to be manufactured. Of these it is worth highlighting the proposal of F. Lalezari, who invented a broadband antenna with the technique of notching, which can be used in missiles and airplanes provided it does not deviate their aerodynamic profile excessively. This antenna also showed frequency-independent

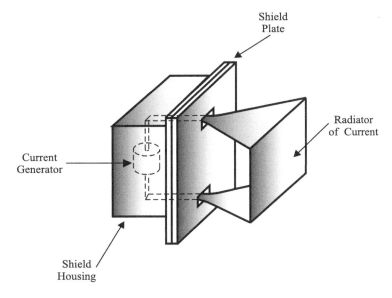

FIGURE 3.4

Design of the current radiator proposed by H. Harmuth in 1985. Adapted from [20].

radiation patterns [18]. Another antenna to mention from this decade is the proposal of M. Thomas, patented under the name of *Wideband arrayable planar radiator*, which was useful in radars and electronic warfare use due to its simplicity, low profile by maintaining a low radar cross section and having polarization diversity. Of both designs, Thomas' antenna has a better performance [7, 19].

Finally, concluding the background to UWB antennas, it is worth mentioning that in 1985 H. Harmuth improved the performance of magnetic antennas by introducing the concept of a current radiator like that shown in Figure 3.4. This magnetic antenna radiates the electromagnetic waves efficiently and with low distortion. This antenna is especially useful to radiate energy in the form of electromagnetic pulses [7, 20].

As already mentioned, since the 1990s various efforts have been made by the scientific community to design non-volumetric UWB antennas, which show a reduction in weight and dimensions. These types of antennas are called planar, whereby from a volumetric structure, it is possible to design a new antenna whose thickness is several times smaller than the other dimensions (i.e., practically a flat antenna). Thus, one could consider two possibilities: (1) an antenna printed on a grounded substrate and (2) an antenna whose radiator is not printed on a grounded substrate (see Figure 3.5). In order to make a

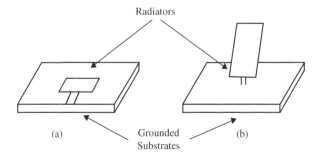

FIGURE 3.5
(a) Planarized antenna (b) planar antenna.

distinction between them, here we call the former *planarized* and the latter *planar*.[2] Let us then begin to examine some aspects of planarized UWB antennas.

3.3 Planarized UWB Antennas

In the past, a serious limitation of antennas with a planarized feeder was their reduced bandwidth, which was between 1 and 10% (compared to a bandwidth from 15 to 50% that is commonly found in the elements of antennas such as dipoles and wave guide horns). However, this limitation was addressed by obtaining an impedance matching of up to 90%, and a gain in bandwidth of up to 70% in separate antennas. To increase the impedance matching of an antenna with a planarized feeder, it was necessary to increase its size, height, and feed-point, as well as to develop some matching techniques [21].

Within this classification there are both balanced and non-balanced antennas. However, the scientific community has focused mainly on the development of non-balanced UWB antennas due to the difficulty in the designs of ultra wideband baluns.

For the purposes of an analysis of planarized UWB antennas, those most representative were selected:

- Vivaldi antenna

- Rectangular patch antenna

- CPW-fed planar ultra wideband antenna having a frequency band notch function

---

[2]It is worth noting that some authors do not make this difference at all and use both terms interchangeably.

- Slot antenna based on precooked ceramic

- Planar volcano-smoke slot antenna

- Printed circular disc monopole antenna

- Microstrip slot antenna with fractal tuning stub

- Planar miniature tapered-slot-fed annular slot antenna

- Tulip-shaped monopole antenna

- Balloon-shaped monopole antenna

- Half cut disc UWB antenna

- Planar UWB antenna array

- Octagonal shaped fractal UWB antenna

We are using the original antenna names given by their creators, but as will be seen, all of them are planarized antennas. It is worth noting here (and in Section 3.4 of planar antennas) that we are also presenting the original variables to describe some antenna parameters, which do not necessarily correspond to those used in other chapters of this book.

3.3.1   Vivaldi antenna

The Vivaldi antenna was first introduced by Gibson in 1979 [22] and is the planarized version of a horn antenna that offers a moderate gain in a low cost structure. Its predecessor is an antenna designed by William Nester, which involved a gradual transition between a microstrip transmission line and a slot transmission line [17]. This is an aperiodic antenna, gradually curved so that at different frequencies, different parts of the antenna radiate (in fact its shape is similar of that of Figure 3.1(b) and also offers a soft transition). The soft curve follows an exponential relationship as proposed by Gibson [22]:

$$y = \pm A \exp(px) \tag{3.1}$$

where y is the half separation distance, x is the length parameter, A is a constant and p is a parameter called the *magnification factor* which determines the beamwidth. Figure 3.6 shows an example where $A = 0.125$ and $p = 0.052$ and Figure 3.7 depicts a model of the Vivaldi antenna as it would be drawn on a printed circuit board. Theoretically, this antenna has an infinite bandwidth and its only limitation is its physical dimensions and the complexity of its manufacturing. In general terms, the feeder determines the higher cut-off frequency limit, while the aperture size defines the lower cut-off frequency limit [23].

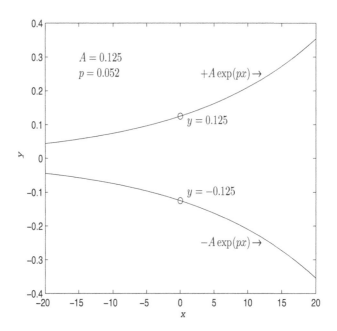

FIGURE 3.6
Example of the exponential expansion used to design a Vivaldi antenna.

In order for the Vivaldi antenna to operate in the microwave region, two crucial factors must be taken into account: (1) the transition from the main transmission line to a line which feeds directly to the antenna should have a very wide bandwidth and a low reflection coefficient; (2) the dimensions and shape of the antenna should be selected according to the purpose of obtaining reduced lateral and back lobes in the whole selected span of frequency [24]. To date, the solutions proposed to supply this balanced antenna are: a balun of ultra wideband or a slotted microstrip transmission line for both sides of the printed circuit [25].

The radiation patterns in the E and H planes of a Vivaldi antenna have a gain greater than 10 dBi, with the half power beamwidth of the main lobe being approximately 60°. However, in some designs this antenna has the disadvantage of presenting a relatively big back lobe, generally in the H plane.

In terms of the reflection coefficient, this type of antenna presents very low matching losses in a frequency range larger than a decade, where the upper and lower cut-off frequencies depend on radiator size. Another characteristic observed from its reflection coefficient is the large quantity of resonance frequencies, many of them at 1 GHz intervals [24].

In addition to the UWB properties of the Vivaldi antenna, other advantages are the directional radiation pattern (with a typical gain of 10

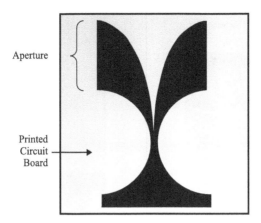

Aperture

Printed
Circuit
Board

FIGURE 3.7
UWB Vivaldi antenna printed on a circuit board.

dBi [24, 25]) and the relatively small and planar size. However, its main limitations are the need for a UWB balun (in fact it is the only balanced antenna considered in this book), since the bandwidth of the balun usually limits the operational bandwidth of the antenna, and the fact that its planar construction limits its use to relatively low power applications. For these reasons, there has been much interest around the world in developing UWB planar monopole antennas with directional characteristics, and of relatively small size that do not require a balun, thus solving the main limitations of this antenna [24].

3.3.2 Planarized antenna with rectangular radiator

This antenna was designed to operate from 3.2 to 12 GHz and it consists essentially of a printed circuit radiator with two graduations at the edges of its lower part and one slot, and a partial ground plane in the rear part of the substrate (see Figure 3.8) [26].

The antenna was built in a perfect electric conductor (PEC) with an area of 15×14.5 mm^2 and printed on an FR4 substrate with a thickness of 1.6 mm and with a relative permittivity of 4.4. The feeder is developed with a microstrip line of $50\,\Omega$. The details of dimensions of this antenna can be found in [26]. The group phase delay is less than 0.5 ns, which is an important parameter that indicates the grade of distortion of the waveform of a pulsing signal that is captured or radiated by the antenna (see Chapter 5 for comments related).

The radiation pattern of the planarized antenna with a rectangular radiator shows quasi-omnidirectional behavior in the XY plane at 3, 5 and 7 GHz. However, this antenna exhibits large pattern variations as it changes with the frequency. As regards the measured and simulated reflection coefficients for this antenna, it is reported that the simulated coefficient is stable and better

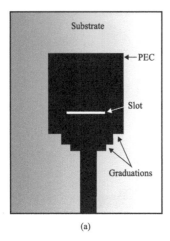 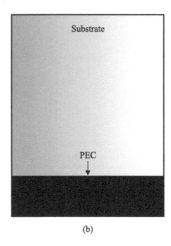

(a)   (b)

FIGURE 3.8
Geometry of the planarized antenna with rectangular radiator: (a) Frontal view, (b) posterior view. (Modified from [26]).

matched than the measured coefficient, which is attributed, in many cases, to mechanical imperfections of the antenna construction.

The planarized antenna with rectangular radiator such as that shown in Figure 3.8, has as a major advantage in its relatively small size and low profile, due to its planar features, but it has as the limitation of an operational bandwidth that is outside the band allocated by the FCC.[3] Another important limitation of this antenna, as already pointed out, is its unstable radiation pattern.

3.3.3   CPW-fed planar ultra wideband antenna having a frequency band notch function

When the FCC assigned the band of frequencies from 3.1 to 10.6 GHz for UWB communications, the possibility of them causing electromagnetic interference near communications systems such as global positioning systems (GPS) or wireless local area networks (WLAN) was considered. For this reason, research has been carried out to propose UWB antenna designs with the characteristic of inserting a guard band at the frequencies of interest. A typical antenna of this type is the antenna with a V-shaped slot proposed in April 2004 [27], which is of a small size and has a guard band centered at 5.25 GHz to avoid interference in WLAN. Kim and Kwon claim that the guard band frequency can be adjusted by varying the length of the V-shaped slot [27].

---

[3]For UWB communications there exists a need of having UWB planarized antennas that have a reflection coefficient magnitude lower than -10 dB and that its operational band is at least inside of the band allocated by the FCC for UWB communications.

Basically, this antenna is printed on a 1 mm thick sheet with an area of 22×31 mm^2 using an FR4 substrate with permittivity $\varepsilon = 4.4$, and the feeder is obtained with a coplanar waveguide (CPW) [27]. This antenna presents a $VSWR < 2$ in the operational frequency band from 2.8 to 10.6 GHz. According to the creators, the introduction of the V-shaped slot produces notches of 10–12 dB magnitudes, which are centered around 5.25 GHz [27]. This antenna also presents a radiation pattern with an average gain of 2.3 dBi in the main lobe direction.

The CPW-fed planar UWB antenna with a frequency band notch function has as major advantages its reduced size, and that its operational band is in line with the frequencies allocated by the FCC for UWB communications. However, the gain in the main lobe direction lower than 2.5 dBi could be considered as a disadvantage.

3.3.4   Slot antenna based on precooked ceramic

Provided that a large part of the current situation of UWB antennas is focused on antenna designs that are useful for portable systems, the slot antenna based on precooked ceramic [28] shows an advantage in comparison with other UWB planarized antenna designs.

This antenna was designed and presented in May 2004 as a solution for UWB systems, since its ground plane can be shared by other radio circuits, thus saving space. This antenna is manufactured in precooked ceramic of a low temperature.[4] The radiant element of the antenna has an ellipse shape with the major axis of 17 mm and the minor axis of 11 mm (see Figure 3.9). The feeder is implemented through a microstrip 41 mm long by 3 mm wide, which provides an input characteristic impedance of 50 Ω. The prototype has a bandwidth from 3.1 to 10.6 GHz with a reflection coefficient smaller or equal to -10 dB [28].

As far as the radiation pattern is concerned, it reports a quasi-omnidirectional pattern at 10.1 GHz, being wider in the horizontal plane. This same behavior is also reported for the radiation pattern at 3.5 and 6.85 GHz [28]. Of its advantages, it is worth highlighting the fact that the slot antenna based on precooked ceramic has a reflection coefficient lower than -10 dB within its operational band (3.1 to 10.6 GHz), which coincides with the band designated by the FCC for UWB communications. In addition, this is a planarized antenna with relatively small dimensions. However, its radiation pattern at the frequencies that were reported exhibits variations larger than 20 dB, compared to an omni-directional pattern.

---

[4]The technology of precooked ceramic of low temperature was developed for the encapsulated microprocessors, however, currently it is used for the development of compact modules of radio frequency and microwaves due to its excellent performance [28].

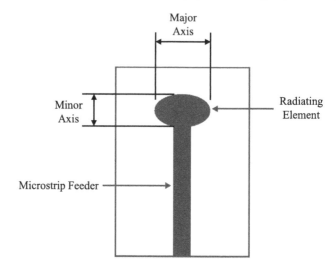

FIGURE 3.9
Slot antenna based on a precooked ceramic technique. (Adapted from [28]).

3.3.5   Volcano-smoke slot antenna

This antenna was proposed by Yeo et al. in August 2004 for possible use in broadband communication wireless applications [29]. It is a planarized antenna with a volcano-smoke-like design (see Figure 3.10). This antenna is made up of a volcano-smoke-shaped patch, an inner island, a ground conductor and a coax-to-CPW transition [29]. This latter feature is crucial to obtain the broadband feature. This type of slot provides a gradual and soft transition from the feeder line to the radiant element.

In order to obtain an input impedance of 50 Ω, Yeo et al. proposed the introduction of a slot in the CPW, which begins with a width of 0.5 mm and is gradually increased until reaching a width in the base of the CPW of 18.5 mm.

In [29] the reflection coefficient of this antenna is reported, which was obtained both experimentally and by means of two different simulation tools. Results show a reflection coefficient magnitude smaller than −10 dB from 0.8 GHz to 7.5 GHz through the Method of Moments, while by using the HFSS (High Frequency Structure Simulator),[5] the result extends from 0.6 GHz up to 7.5 GHz. Finally, the experimental measurement shows an operational bandwidth from 0.8 GHz up to 6.7 GHz for a VSWR lower than 2.3.

In terms of the radiation pattern, a quasi-omnidirectional pattern at 0.8 GHz is reported, which suffers alterations as the frequency increases, presenting nulls of up to 30 dB. However, at 5 GHz, the radiation pattern shows not

---

[5]See Chapter 9

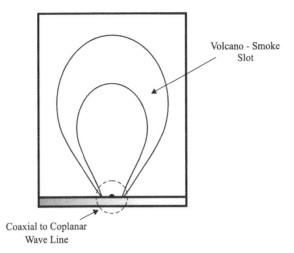

FIGURE 3.10
Geometry of a volcano-smoke slot antenna. (Adapted from [29]).

only nulls, but also large variations [29]. For this reason, authors investigated a way of modifying part of the radiation pattern to make it directional, in order to attenuate the part of the radiation pattern that was considered as lateral lobes. Thus, experiments were carried out inserting a 25 mm width FGM-40 absorber in the antenna rear, with satisfactory results [29].

Like the most representative UWB antennas, this antenna keeps the advantages of having relatively small dimensions and a reflection coefficient magnitude smaller than -10 dB within its operational band. However, in contrast to the previous antenna presented in Section 3.3.4, it has the disadvantage that its operational band coincides only partially with the band allocated by the FCC for UWB communications.

3.3.6 Printed circular disc monopole antenna

This planarized antenna was proposed in September 2004 [30] as an alternative for UWB planar monopoles with a simple structure and easy construction in circular, square, elliptic, pentagonal and hexagonal shapes, and whose radiation and bandwidth properties were satisfactory, but whose integration on printed circuits was not appropriate since, these being latter planar antennas, their ground planes are perpendicular to their radiators.

The planarized antenna with a printed circular disc monopole is fed by a microstrip line, and based on its optimum design, as Liand et al. state, other antennas can be developed to have a very wide bandwidth with quasi-omnidirectional patterns in the whole bandwidth [30].

Just like the simply built monopoles, this antenna reported that the operational band critically depends on separation, as well as the width of the ground plane W, hence the optimization of these two parameters provided the maximum bandwidth [30]. During the design of this antenna it was found that the optimum separation between the feed-point of the radiator and the ground plane is $h = 0.3$ mm, while the optimum width of the ground plane is $W = 42$ mm [30].

From the simulated and measured reflection coefficients of the printed circular disc monopole antenna, it was reported that the measured bandwidth of this antenna is from 2.78 to 9.78 GHz, while the simulated bandwidth is from 2.69 to 10.16 GHz [30]. Although this antenna does not completely cover the bandwidth designated by the FCC, it definitely presents UWB characteristics.

The radiation pattern, on the other hand, is shown as quasi-omnidirectional in the H plane for the whole operational frequency band. Additionally, the convergence between the measured and simulated radiation patterns is an aspect highlighted in [30].

Like other antennas presented up to this point, the printed circular disc monopole antenna has among its main advantages the fact that it is a planarized antenna with a reflection coefficient magnitude smaller than −10 dB within its operational band, and that it has dimensions similar to those of other designs. However, among its main disadvantages is the fact that it has an operational band of only 7 GHz, and that although the inferior cut-off frequency of its operational band is smaller than that allocated by the FCC for UWB communications (2.78 GHz instead of 3.1 GHz), it does not cover the whole frequency band, since its bandwidth is up to 9.78 GHz instead of 10.6 GHz. Additionally, no information is available about the frequency response of the radiation pattern, gain, and phase linearity, among other features.

3.3.7   Microstrip slot antenna with fractal tuning stub

This compact-dimension antenna, proposed in March 2005 by Lui et al. [31], was designed starting from a conventional slot patch antenna, to which a fractal tuning stub was added in order to have a guard band between 4.95 and 5.85 GHz, to avoid interferences with communications systems near to WLAN (i.e., in the same way as the CPW-fed UWB antenna addressed in Section 3.3.3, although for a wider guard band). The operational bandwidth obtained with this antenna is from 2.66 to 10.76 GHz, which is useful in UWB communications systems. Its omnidirectional radiation patterns are quasi-stable, presenting an efficient performance in its whole operational band.

The reflection coefficient magnitude reported for this antenna is −10 dB for the whole operational bandwidth, with the exception of the guard band located at approximately 5 GHz. As far as the inferior cut-off frequency is concerned, the result obtained by simulation showed a value lower than the measured frequency. In contrast, the upper limit shows behavior similar to the simulated result.

With regard to the radiation pattern, results are presented in [31] at three different frequencies: 3, 7, and 10 GHz. The radiation pattern has a quasi-omnidirectional behavior for the whole operational bandwidth. However, as with other antennas considered in this chapter, the radiation patterns of this antenna are, as expected, frequency dependent [31].

The microstrip slot antenna with fractal tuning stub presents among its main advantages a reduced size, a reflection coefficient magnitude smaller than −10 dB in its operational band, a fractal tuner for a guard band, and an operational band in keeping with the spectrum authorized by FCC for the UWB communications. Although the radiation patterns are quasi-omnidirectional, they present an irregularity of around 20 dB for the frequencies presented.

3.3.8 Planar miniature tapered-slot-fed annular slot antenna

The design of this antenna was proposed by Ma and Jeng in March 2005, being the only one with an annular slot feeder structure that has a $VSWR < 2$ and a quasi-omnidirectional and quasi-stable radiation pattern in the whole band of frequencies allocated by the FCC for UWB applications [32].

The geometry of this antenna can be observed in Figure 3.11. As can be seen the radiator is built in the frontal side of the substrate, while the microstrip line and its open stub were built in its rear side. The substrate has a thickness $t = 1.57\,\text{mm}$ and a relative permittivity of 2.2. Ma and Jeng explain that the energy in this antenna is transferred first from the microstrip line to the slot through a broadband transition. This tapered-slot feeder carries out the function of an impedance transformer and guides the wave toward the radiation slot without causing important reflection. Finally, authors assert that the radiation slot is curved in order to distribute part of the energy to the opposite side of the aperture feeder [32].

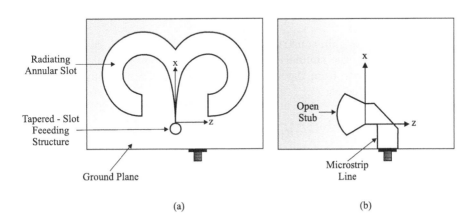

(a) (b)

FIGURE 3.11
Geometry of a planar miniature tapered-slot-fed annular slot antenna: (a) Frontal face, (b) posterior face. (Modified from [32]).

From the design considerations presented by Ma and Jeng, two prototypes were manufactured, one miniaturized prototype and another as a reference, with the final dimensions of the former being $35.6 \times 40.3 \,\text{mm}^2$ and the latter $46.5 \times 66.3 \,\text{mm}^2$ [32]. From the measurements obtained from the two prototypes, the reported reflection coefficient establishes an operational frequency band for the reference antenna (with a $VSWR < 2$) that covers the spectrum of UWB almost totally, whereas the miniature antenna offers an operational frequency band from only 4.8 to 10.2 GHz [32].

With regard to the radiation pattern, this is only reported for the reference antenna at 6.5 GHz in the E plane and the H plane. The radiation pattern in the H plane is uniform, but the radiation pattern in the E plane exhibits properties of dual polarization. The gain of the reference antenna is from 4 to 6 dBi in the band from 3.1 to 9 GHz [32].

The miniature tapered-slot-fed annular slot antenna has advantages similar to the antennas previously analyzed; however, it presents the disadvantage that although the reference antenna for the development of this antenna has an operational band that covers almost the whole band of frequencies allocated by the FCC for UWB communications, the miniaturized antenna does not.

3.3.9 Tulip-shaped monopole antenna

This type of antenna, proposed in 2006, has a tulip-shaped structure. Chang et al. justify the use of a petal radiator as the basis on which to provide a quasi-omnidirectional radiation pattern in a frequency band wider than that established by the FCC for UWB [33]. The main structure of this antenna is depicted in Figure 3.12, where the tulip shape is easily observed. This structure can be seen in two parts: a half-ring base with a radius of R_1 and a corolla, whose inner curve is formed from a radial section of a circle with a radius of R_4.

There are two main construction features of this antenna, given by Chang et al. [33]. First, the simulation of the half-ring mode current distribution, whereby they state from the simulated results that the resonance frequency can be represented as:

$$f_s = \frac{nc}{\pi R_1 \sqrt{\varepsilon_{eff}}} \quad \forall \quad n = 1, 2, \ldots \quad (3.2)$$

with ε_{eff} being the effective dielectric constant and c the light speed in free space. So, the rest of the tulip-shaped structure is completed by simulating both the calyx-mode and the corolla-mode current distributions until achieving acceptable results.

The technical specifications of the design of this prototype are: $R_1 = 12.7 \,\text{mm}$, $R_2 = 0.25 R_1$ (3.175 mm), $R_3 = 0.5 R_1$ (6.35 mm) and $R_4 = 4/3 R_1$ (16.93 mm). The substrate is an FR4 with a width of 0.8 mm and a relative permittivity of 4.4 [33].

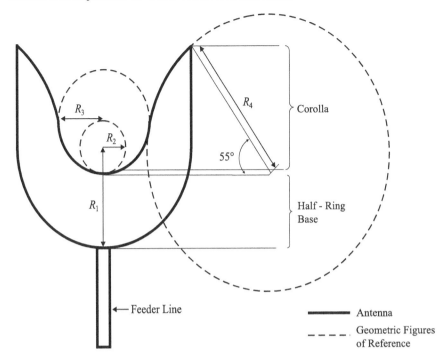

FIGURE 3.12

Configuration of the tulip-shaped monopole antenna. (Modified from [33]).

An important result reported by Chang et al. is that this antenna presents an impedance matching with a relation of 1:16 [33], which is not common among UWB antennas, where there is usually a bandwidth ratio of less than 1:10 (in fact, the authors make reference to only two antenna designs with a bandwidth ratio higher than 1:10). Thus, through an appropriate tuning of the tulip-shaped structure, the measured reflection coefficient magnitude reported for this antenna shows a value smaller than −10 dB from 2.55 to 32.5 GHz, and hence the operating band allocated by the FCC is entirely covered by this design, so it could be useful for spectrum monitoring purposes.

In terms of radiation patterns, the authors report results at 3.1, 7.1, and 10.6 GHz in the planes XY, XZ and YZ. The behavior of this radiation pattern obtained by simulation is the same as that of a typical monopole antenna, especially in the quasi-omnidirectional radiation pattern of the XZ plane. For the measured pattern, this behavior is only kept at 3.1 GHz, while at 7.1 and 10.6 GHz some nulls become present at the same XZ plane. With regard to the gain, a value of 0.2 dBi is reported at 3 GHz, reaching up to 4 dBi at 10.5 GHz.

The tulip-shaped monopole antenna is one of the most advantageous antennas, due to its reduced size (total area of $33.4 \times 37.1 \text{ mm}^2$), the fact that it covers, as already mentioned, the band of frequencies allocated by the FCC for UWB communications, and its relatively simple construction.

The relatively small dimensions of this antenna impose certain constraints on transmission at low power levels. If directive radiation patterns are needed for a specific application, this antenna would be also limited. Just as with other designs, the tulip-shaped antenna introduces irregularities in the radiation pattern (up to 20 dB) as the frequency increases (in fact, the radiation pattern for frequencies higher than 10.6 GHz are not presented, so it is difficult to analyze this parameter in the whole operation band proposed by the authors for this antenna). All of the above could be considered as disadvantages, but they are not limitations.

3.3.10   Balloon-shaped monopole antenna

Another application of UWB antennas is for radio-frequency identification (RFID), where the scattering characteristics become a crucial aspect to take into account. In this context, Hu et al. designed in 2008 a UWB balloon-shaped monopole antenna while looking for the requirements of the RFID tags, particularly of the passive type, which means that the antenna design should be considered without a battery and chip [34].

Provided authors designed this UWB-RFID antenna for a passive tag operation, they proposed variable antenna terminations in such a way that all possible UWB pulses of comparable amplitude can be backscattered. To achieve that, this antenna has a feed line with a different length, which allows controlling the time interval between UWB pulses.

The antenna size is compact since the balloon has only a 18 mm minor axis and 22 mm major axis, and the board over which the whole antenna is printed (balloon shaped radiator plus feed line) is $23 \times 31 \, \text{mm}^2$. This device was manufactured using a Rogers RO4003C substrate 0.508 mm thick. With all the aforementioned features (and specifications of the substrate and all dimensions detailed in [34] of course) this antenna provides a simulated and measured impedance bandwidth from 3.6 to 12.4 GHz and from 3.4 to 11.3 GHz, respectively.

This balloon-shaped monopole achieves a quasi-flat gain response (around 4.7 dBi between 3 and 9 GHz approximately) and, although the radiation pattern is not graphically presented in the paper, the authors claim that an almost omnidirectional pattern is achieved.

Regarding the scattering results, the authors briefly present definitions derived from the Radar Cross Section (RCS): what is known as *structural mode* and *antenna mode*, where the former corresponds to "the early-time pulse which is scattered from a matched-loaded antenna" and the latter is related to "the late-time response which is due to the termination mismatching and reradiation" [34]. From these definitions, derived from a theoretical framework and their observations, it is concluded that this antenna is suitable for passive RFID applications.

3.3.11   Half cut disc UWB antenna

In the category of UWB planarized antennas, those designed to be used for wireless USB (Universal Serial Bus) applications have recently gained interest in the research community. In this context, one of the critical aspects is definitely the size of the antenna, which should be as compact as possible so that it can be integrated to the USB toggle. One of the works addressing this problem is that presented by Liu and Chun in 2009 [35], who explored how to reduce the size of the well known UWB disc (see Section 3.3.6 for instance).

Basically, their proposal is to halve the antenna symmetrically, which, obviously reduces its size, but a mismatching is introduced. In order to resolve this problem, the authors first apply a beveling technique to the ground plane in such a way that a smooth transition from one resonant mode to another is accomplished. Second, a tapered CPW fed strip and gradient gap are also implemented. Thus, a good match is achieved. The entire dimension of the antenna is 11 mm × 29.3 mm × 1.6 mm. The CPW transmission line was designed for a subminiature A (SMA) connector. Another important factor is that both the radiator and the feeding structure were implemented on the same plane, which allows single-sided metal on only one layer of the substrate and hence a low manufacturing cost. The impedance bandwidth is from 2.42 to 13.62 GHz in measurement, which covers the entire UWB band. Regarding the radiation pattern, it is quasi-omnidirectional with a stable gain from 1.67 to 5.88 dB over the UWB frequency range. Therefore, this tapered CPW half cut disc antenna seems to be a good candidate for wireless USB implementation.

3.3.12   Planar UWB antenna array

For several years there has been interest in the use of antennas for medical applications. UWB antennas are no exception, and an example of this is the antenna developed by Sugitani et al. in 2012 for breast cancer detection [36]. Although this is not a single element (it is an antenna array), it can be considered in the category of planarized UWB antennas in this section provided the slots that make up this array are printed on the same board. Thus, the main objective of the authors was to search for an antenna that could be smaller than the others in the open literature in those years and still conserve the detection capabilities. In this direction, they presented a square 4 × 4 elements antenna array of dimension 44 × 52.4 mm^2 which presents an impedance matching between 3.5–15 GHz. No information is provided about the radiation pattern, but according to the authors' measurements, this antenna can perform well for cancer detection, since up to two human breast tumors located at the depth of 20 mm with 20 mm of separation could be detected. Details of how they simulate the body conditions will be described in Chapter 8.

3.3.13   Octagonal shaped fractal UWB antenna

It is worth concluding this section with a very recent design: the octagonal shaped fractal UWB antenna presented in 2013 [37]. According to the authors, the central idea that motivated this structure is the use of fractal theory as a mechanism to reduce the whole antenna size. However, it also caused a bandwidth reduction, for which techniques such as feed gap and insertion of multiple notches in the ground plane were implemented, in such a way that the impedance bandwidth covers the UWB range.

Then, a Rogers DT/Duroid 5880 substrate of $13.5 \times 16.5 \, \text{mm}^2$ size with a dielectric constant of 2.2 was chosen, over which a combination of the known Monkowski-like fractal and an octagonal basis was printed. On the posterior side of the substrate the notched ground plane was printed. Under these conditions including details given in [37] an almost omnidirectional radiation pattern is achieved for a couple of frequencies within of the UWB band (4.2 GHz and 9.4 GHz), from where it could be preliminary considered that could present a stable radiation pattern through the band of interest.

From the results reported, this antenna presents the advantages of size reduction, good performance in terms of return loss and radiation pattern, and a relatively low cost. As Tripathi et al. pointed out, its omnidirectional radiation feature provides it the possibility to be a candidate for many applications related to short range wireless networks like body area networks.

3.4   UWB Planar Monopoles Antennas

This type of antenna is the result of research tending to satisfy the need for UWB antennas with omnidirectional radiation patterns for military and civilian applications. The classical solution to obtaining an omnidirectional radiation pattern is to use a dipole or a monopole with a ground plane. However, the dipole and the wire monopole have the disadvantage of a narrow band nature. The bandwidth can be increased using a flat metallic structure instead of a thin wire [2]. In fact, the exploration of the behavior of diverse flat metallic radiators with different structures is the foundation of UWB planar monopole antennas. Currently, these antennas have provoked interest in the scientific community, again because they have a simple geometry, and easy construction. Thus, four designs have been selected as the most representative of the current situation of UWB planar antennas:

- Planar inverted cone antenna (PICA)

- Bi-arm rolled monopole antenna

- Square planar monopole antenna with notched technique

- Planar directional monopole antenna with leaf form

- Compact UWB antenna

3.4.1 Planar inverted cone antenna (PICA)

This antenna was proposed in 2004 by Suh, Stutzman, and Davids as an alternative to the degradation of the radiation pattern suffered by other designs at the high end of their impedance matching [38]. Basically, this antenna is made up of an inverted cone-shaped flat radiant element, which is mounted on a perpendicular ground plane as shown in Figure 3.13. The authors claim that although the geometry of this antenna is very simple, it provides outstanding performance in terms of impedance matching and radiation pattern with a bandwidth larger than a decade.

The general geometry of the PICA shown in Figure 3.13 converges to the basic structure when the dimension $W_1 = 0$. By varying the dimension W_1, both the shape and size of the antenna are modified, providing the mechanisms to achieve certain performance for specific applications. The shape of W_2 affects the bandwidth and radiation pattern, although the authors point out that the circular, elliptical, and exponential shapes provide quasi-omnidirectional radiation patterns and an ultra wideband [38].

In addition, Suh et al. present the possibility of expanding the operational band of the PICAs, which consists of simply adding two circular holes in the

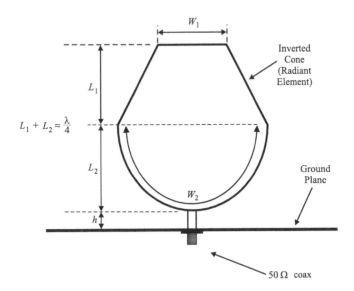

FIGURE 3.13
General geometry of the PICA. (Adapted from [38].)

radiant element. According to the results given in [38], the addition of the two perforations in the radiator element improves the impedance matching at the higher part of the band achieving a $VSWR < 2$ in a band wider than a decade, without significantly altering the radiation pattern. By inspecting the radiation pattern of a PICA with and without perforations as presented by the authors, one can observe a depression of more than 30 dB at a frequency of 1 GHz, which is decreased to approximately 20 dB at 7 GHz, although the pattern becomes more irregular. However, relatively fewer variations can be seen in the radiation pattern of the two circular holes in the PICA antenna.

The reported radiation pattern is measured at 1, 3.4, and 7 GHz. This radiation pattern is gradually degraded as the frequency is increased, such that Suh et al. consider that depending on the application, this degradation can provide the upper cut-off point. The gain also increases with frequency, from 5 dBi at 1 GHz to 8 dBi at 7 and 10 GHz [38].

Among the main advantages of the planar inverted cone antenna are its higher gain levels compared to those reported by other designs (e.g., [33]). It has also the capacity to radiate higher power levels than those provided by a planarized antenna. In addition, its radiator is relatively small for an operational frequency band of a decade from 1 to 10 GHz. However, this antenna has the following disadvantages: the ground plane is relatively large in comparison with the radiator (a relation of almost 1:10); its operational frequency band does not comply with the frequencies allocated by the FCC for UWB communications, given that this is only reported up to 10 GHz and not up to 10.6 GHz (the radiation pattern at 10 GHz is not reported either).The above can be attributed to the increase in variations of the pattern at high frequencies.

3.4.2   Bi-arm rolled monopole antenna

This is a continuation of the research efforts targeted at the planar monopoles, since these are of a small size, low cost and wide band. However, they have the disadvantage that at their high operating frequencies, their radiation patterns tend to be directive due to their asymmetrical structure. As such, this antenna was proposed in 2005, which has an omnidirectional and broadband radiation pattern. The bi-arm rolled monopole antenna is formed by rolling a planar monopole symmetrically in relation to its half part (midline). The characteristic impedance and the radiation pattern were investigated experimentally and they were compared with a planar rectangular monopole and a narrow strip monopole. In addition, the transfer function of an antenna system formed by a pair of identical monopoles is presented [39].

The bi-arm rolled monopole antenna has a height of 16 mm and a maximum diameter of 3 mm. The rolled monopole was manufactured by symmetrically rolling a rectangular copper sheet of $13 \times 16 \, mm^2$. This generates a pair of symmetrical arms with a semicircular section and a planar interior, as can be seen in Figure 3.14. For comparison purposes, prototypes of a planar

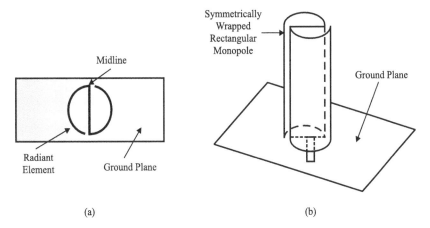

(a)   (b)

FIGURE 3.14
Geometry of the bi-arm rolled monopole. (a) Top view, (b) Three-dimensional view.

rectangular monopole of $13 \times 16\,\text{mm}^2$ and a narrow strip monopole of $3 \times 16\,\text{mm}^2$ were also built by the authors.

Of the three monopole pairs, the rectangular monopole has the widest bandwidth (3.4 to 10.2 GHz). The bi-arm rolled monopoles have an impedance similar to the rectangular monopoles, but their impedance matching is achieved in a smaller bandwidth (5.3 to 9.5 GHz), while an impedance matching between 3.8 to 5.3 GHz was obtained from the narrow strip monopole, [39].

Compared with the flat rectangular monopole, the bi-arm rolled monopole antenna has the advantage of having a radiation pattern that is perfectly omnidirectional in the horizontal plane [39], thus, it can be considered as a good option for applications of UWB wireless communications where this feature is required.

Among the main advantages of the bi-arm rolled monopole antenna is that it has the capacity to transmit power levels that are higher than the planarized antennas, its radiator is the smallest of all the antennas included in this chapter, and it has a radiation pattern that is perfectly omnidirectional in the horizontal plane. However, this antenna does not completely cover the FCC allocated band for UWB communications, since it only has an impedance matching from 5.3 to 9.5 GHz.

3.4.3 Square planar monopole antenna with notching technique

The square planar monopole antennas have a simple geometry, and are easy to construct starting from a simple metal board. However, as previously mentioned, this type of antenna has the disadvantage of having a small bandwidth.

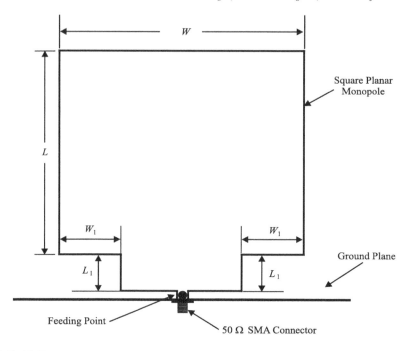

FIGURE 3.15
Geometry of the square planar monopole antenna of UWB (modified from [40]).

In order to improve the impedance matching into a wider bandwidth and to make this kind of antenna useful for wireless local area communications, in September 2004 a square planar monopole antenna with a notching technique was proposed [40]. By using this technique the two inferior corners of the monopole are cut in an appropriate dimension, increasing the impedance matching by four times compared to a conventional square flat monopole (up to approximately 10.7 GHz from the original 2.5 GHz).

Figure 3.15 shows the square planar monopole antenna with two cuts in its lower corners. This square monopole antenna with the dimensions $W \times L$ is manufactured starting with a metal copper board 0.2 mm thick, which is mounted above a ground plane of 100×100 mm^2. In the center of the monopole bottom, a stub 2 mm wide and with length h protrudes at point A (the feedpoint) through a hole in the ground plane, to be attached to an SMA connector of 50 Ω [40].

For a conventional square planar monopole (where $W_1 = L_1 = 0$ in Figure 3.15), the maximum impedance matching is achieved by adjusting the value of h (the optimal value of h is between 2.5 and 3.0 mm [40]) for an L that varies from 25 to 55 mm. The bandwidth that is obtained with this optimal value is from 1 to 3 GHz, depending on the size of the planar monopole. Su et al. assert that by making the previously mentioned cuts to the appropriate dimensions ($W_1 \times L_1$), an improvement in the impedance matching is achieved for this type of antenna, and that this behavior is attributed mainly to the fact that these cuts influence the matching between the planar monopole and the ground plane [40].

The simulated and measured reflection coefficient magnitude reported for a square planar monopole antenna with $L = 30$ mm, $W_1 = 7$ mm, $L_1 = 3$ mm and $h = 1.5$ mm presents an impedance matching smaller than -10 dB for a frequency band from 1.9 to 12.7 GHz. It is worth pointing out that with the selection of the dimensions $L = 30$ mm and $h = 1.5$ mm, the inferior cut-off frequency obtained is below 2 GHz. The optimal parameters W_1, L_1 and h were determined experimentally and they were confirmed later on with a simulation tool [40].

From the measured and simulated radiation patterns reported at 7.5 GHz, the results show a good match between the simulated and measured values [40]. As in previous cases, the radiation pattern depends significantly on the operating frequency.

In order to improve the response of the radiation pattern as a function of the frequency, especially in the azimuthal plane (XY plane), Su et al. used the two planar orthogonal monopole method, claiming that it can improve the omnidirectional characteristics of the square planar monopole antenna [40]. This also impacts on the gain, which increases steadily from 2.8 to 8.0 dBi, as the operating frequency of the antenna is increased [40].

As with the previous monopoles, the main advantages of the square planar monopole antenna with notching technique include its ability to transmit more power than planarized antennas, and its simply built radiator, which allows an impedance matching smaller than -10 dB from 2 to 11 GHz, thus covering the band allocated by the FCC for ultra wideband wireless communications. Finally, it has an omnidirectional radiation pattern with a gain between 2.8 and 8 dBi.

3.4.4 Planar directional monopole antenna with leaf form

The design of this antenna presented by Yao et al. in [41], is based on the fact that in order to improve the bandwidth of a microstrip antenna, one of the most practical solutions is simply to enlarge the substrate using materials of low permittivity like air, as long as the manufacturing costs are not increased. Thus, the authors propose three elements on which their designs are based [41]: a leaf-form radiator, a ground plane and an air substrate between radiator and the ground plane when the former is inclined at a certain angle $\alpha \neq 0$ (see Figure 3.16).

The leaf-form radiator with dimensions $W \times L$ follows a square cosine law at its bottom edge. The minimal and maximal distances between the radiator and the ground plane are $h_{\min}$ and $h_{\max}$, respectively, with $h_{\min} = 1$ mm and $h_{\max}$

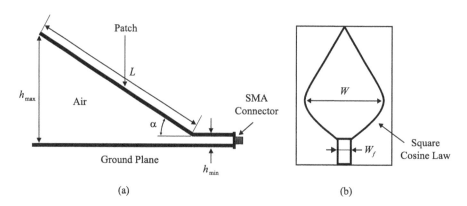

FIGURE 3.16
Design geometry of the planar directional monopole antenna with leaf form.
(a) Lateral view, (b) frontal view. (Modified from [41]).

being approximately half of the wavelength of the inferior cut-off frequency
of the operational frequency band. The angle α between the radiator and the
ground plane can be varied to obtain an optimal performance. A microstrip
line feeder is connected at the bottom of the radiator, which is parallel to the
ground plane and centered at the centerline of the radiator. The wide w_f of
the microstrip line is designed in such a way (4.91 mm) that it achieves an
impedance of 50 Ω [41].

The reflection coefficient magnitude reported for this antenna with an optimal angle of $\alpha = 30°$, presents an impedance matching smaller than -10
dB for a frequency span from 3.05 to 26.87 GHz, which represents a relation
of 8.8:1 [41]. As regards the radiation pattern, a measured pattern at 6 GHz
and 18 GHz is reported where a narrow beamwidth and relatively stable lobule is presented in the E plane, whereas the radiation pattern is less stable
in the H plane. Among the main advantages of the leaf-shaped planar directional monopole antenna are its reduced dimensions, an operational band that
covers the operating band allocated by the FCC for UWB operations, and a
directional radiation pattern that is almost constant through the frequency
band of interest.

3.4.5 Compact UWB antenna

A relatively recent application of UWB antennas is for Body Area Networks
(which will be explained in Chapter 8), where low profile omnidirectional
antennas with high time domain resolution are required. Several studies have

been carried out in the matter; from these, let us take the work of Chahat et al. [42] as an example to explain the antenna characteristics and capabilities. This paper is focused particularly to on-body BAN, where devices that made up the network communicate with each other for a specific application (e.g., health care) and therefore the propagation conditions on the body surface have to be as good as possible.

After early simulations of the impedance matching in free space for different slot configurations and ground plane sizes, the antenna resulted in a compact microstrip-fed printed monopole of $25 \times 10 \times 1.6\,\mathrm{mm}^3$ printed on a $1.6\,\mathrm{mm}$ thick AR350 substrate with dielectric constant of 3.5 (see Figure 1a of [42]). An important characteristic of this antenna is the type of polarization (electric field), which is perpendicular to the body surface. As the authors explain, this polarization allows better propagation conditions for on-body links. Thus, posterior simulations and measurements were carried out using a phantom, whose purpose is to introduce the dielectric characteristics of the body (authors were specially interested in arm muscle) and over which the compact antenna was perpendicularly mounted.

With all the aforementioned dimensions, substrate, and antenna polarization, simulations and measurements were conducted using the phantom whose results reported are as follows. The simulation and measurements for the reflection coefficient are in relatively good agreement and show an impedance matching almost covering the 3.1–10.6 GHz range for UWB systems. As for the radiation pattern, it exhibits a quasi-omnidirectional shape, which is lightly altered at the higher frequency (10 GHz).

3.5   Double-Ridged Guide Horn Antenna

As mentioned in the Section 3.1, this antenna is discussed in a separate section, since it can be used as a standard antenna to characterize other antennas.

The typical standard antennas suggested by the ANSI standard C63.4 for frequencies higher than 1 GHz are the *standard gain horns*. Moreover, as Botello-Perez et al. comment [43], the last version of the standard Mil-Std-461-E establishes that one of the antennas to use in the range from 1 to 18 GHz is the double ridged guide horn (DRGH) antenna. Among the most important characteristics of this type of antenna is that the DRGH can operate with comparatively high power levels, the characteristics of its radiation pattern remain almost constant in a large part of its operating band, its reflection coefficient is small, and its construction is relatively simple [44].

For this horn antenna (see Figure 3.17) to be able to operate in a wide frequency range, two requirements should be satisfied: first, the characteristics of its radiation pattern in the whole operating band must be kept constant; second, this antenna can be fed by a waveguide and at the same time keep its

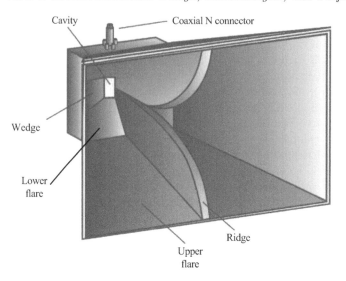

FIGURE 3.17
Geometry of the design of the double-ridged horn antenna.

input impedance constant. The constancy of the radiation pattern is determined by both the size of the antenna aperture and the phase deviation of the electromagnetic field through the aperture. The aperture size is a function of the width and height of the mouth of the pyramidal section. The larger the aperture, the greater the gain, up to the point where the phase deviation is such that the gain begins to diminish. The phase deviation is a function of both the expansion angle and the length of the pyramidal section, so the phase deviation is small when the pyramid is very long and the expansion angle is small, or when the pyramid is very short and the expansion angle is very large [44].

The antenna developed in [44] shows a $VSWR < 2$ in the frequency span from 2.5 to 16 GHz, thus the measurements of radiation patterns of the different designs to be addressed in Chapters 6 and 7 will be limited to this range of frequencies, since this antenna is used as a reference.

3.6   Comparison of UWB Antennas

In this section, a comparison of the different types of UWB antennas is presented. Tables 3.1 and 3.2 show the most important characteristics of these antennas in order to compare them [23].

TABLE 3.1

Comparative table of UWB antennas (Part 1).

Antenna type	Gain (dBi)	Feeder[†]	Dimensions
Vivaldi antenna	≈ 10	Bal.	$150 \times 120\,\text{mm}^2$
Rectangular patch antenna	< 5	Unbal.	$15 \times 14.5\,\text{mm}^2$
CPW-fed planar UWB antenna with notch function	< 5	Unbal.	$22 \times 31\,\text{mm}^2$
Slot antenna based on precooked ceramic	–	Unbal.	$11 \times 17\,\text{mm}^2$
Planar volcano-smoke slot antenna	–	Unbal.	$130 \times 130\,\text{mm}^2$
Printed circular disc monopole antenna	–	Unbal.	$42 \times 50\,\text{mm}^2$
Microstrip slot antenna with fractal tuning stub	< 5	Unbal.	$48 \times 41\,\text{mm}^2$
Planar miniature tapered-slot-fed annular slot antenna	$4 - 6$	Unbal.	$46.5 \times 66.3\,\text{mm}^2$
Tulip-shaped monopole	$0.2 - 4$	Unbal.	$33.4 \times 37.1\,\text{mm}^2$
Balloon-shaped monopole	≈ 4.7	Unbal.	$23 \times 31\,\text{mm}^2$
Half cut disc UWB antenna	$1.6 - 5.8$	Unbal.	$11 \times 29.3\,\text{mm}^2$
Planar UWB antenna array	–	–	$44 \times 52.4\,\text{mm}^2$
Octagonal shaped fractal UWB antenna	–	Unbal.	$16.5 \times 13.5\,\text{mm}^2$
Planar inverted cone antenna	< 9	Unbal.	$76.2 \times 76.2\,\text{mm}^2$
Bi-arm rolled monopole	< 9	Unbal.	$16 \times 3\,\text{mm}^2$
Square planar monopole with notching technique	$4 - 6$	Unbal.	$30 \times 30\,\text{mm}^2$
Planar directional monopole antenna with leaf form	–	Unbal.	–
Compact UWB antenna	$1 - 1.6$	Unbal.	$25 \times 10\,\text{mm}^2$
Double ridged gain horn	$11 - 16$	MP	$240 \times 142 \times 152\,\text{mm}^3$

[†] Bal: Balanced. Unbal: Unbalanced. MP: Magnetic Probe

In summary, the efforts of the scientific community have been concentrated on unbalanced antennas, since the manufacturing of a broadband balun represents additional complications for the design of a UWB antenna. On the other hand, a marked interest is noted in omni-directional antennas of reduced dimensions (mainly of slot and microstrip) rather than directional designs, due to the necessity of providing antennas for the wireless mobile communication systems in the band allocated by the FCC for UWB.

TABLE 3.2
Comparative table of UWB antennas (Part 2).

Antenna type	Bandwidth (GHz)	Comments
Vivaldi antenna	Unlimited[†]	Wideband balun
Rectangular patch antenna	3.2 – 12	
CPW-fed planar UWB antenna with notch function	2.8 – 10.6	Guard band 5.15 – 5.35 GHz
Slot antenna based on precooked ceramic	3 – 10.6	LTCC technology
Planar volcano-smoke slot antenna	0.8 – 7.5	
Printed circular disc monopole antenna	2.78 – 9.78	
Microstrip slot antenna with fractal tuning stub	2.66 – 10.76	Guard band 4.95 – 5.85 GHz
Planar miniature tapered-slot-fed annular slot antenna	3.1 – 10.6	
Tulip-shaped monopole	2.5 – 40.5	The biggest bandwidth
Balloon-shaped monopole	3.4 – 11.3	
Half cut disc UWB antenna	2.42 – 13.62	
Planar UWB antenna array	3.5 – 15	
Octagonal shaped fractal UWB antenna	3.1 – 10.6	
Planar inverted cone antenna	1 – 10	
Bi-arm rolled monopole	3.1 – 10.6	The smallest antenna
Square planar monopole with notching technique	1.979 – 12.738	
Planar directional monopole antenna with leaf form	3.05 – 26.87	Directional pattern
Compact UWB antenna	≈ 3.1 – 10.6	
Double ridged gain horn	2.5 – 16	Standard antenna

[†] Theoretically.

3.7   Conclusions

The study of recent developments in UWB antennas allowed us to identify the fact that the efforts of the scientific community are centered on planar or planarized antennas, that allow high mobility and portability, with special interest in those that have a stable omnidirectional radiation pattern, mainly in the band of frequencies allocated by the FCC for UWB communications.

In order to be able to tackle the design of both omnidirectional and directional planar and planarized monopole UWB antenna, with a simple construction, the theory associated with UWB antennas is presented in

Chapter 4, where in addition to diverse definitions and parameters related to such designs, several equations proposed by different authors for the development of UWB antennas are given.

Bibliography

[1] H. Schantz. Introduction to ultra-wideband antennas. In IEEE, editor, *EEE Conference on Ultra Wideband Systems and Technologies*, 2003.

[2] W. L. Stutzman and G. A. Thiele. *Antenna Theory and Design*. John Wiley & Sons, 1998.

[3] J. D. Kraus and R. J. Marhefka. *Antennas for all Applications*. McGraw-Hill, 2002.

[4] FCC. First report and order, revision of part 15 of the commission's rules regarding ultra-wideband transmission systems. Technical report, Federal Communications Commission, 2002.

[5] X. H. Wu, Z. N. Chen, and M. Y. W. Chia. Note on antenna design in UWB wireless communication systems. In *2003 IEEE Conference on Ultra Wideband Systems and Technologies*, pages 503–507, 2003.

[6] O. J. Lodge. Electric telegraphy, 1898.

[7] H. Schantz. A brief history of UWB antennas. *IEEE Aerospace and Electronic Systems Magazine*, 19(4):22–26, 2004.

[8] P. S. Carter. Short wave antenna, 1939.

[9] P. S. Carter. Wide band, short wave antenna and transmission line system, 1939.

[10] S. A. Schelkunoff. Ultra short wave radio system, 1941.

[11] N. E. Lindemblad. Wide band antenna, 1941.

[12] L. N. Brillouin. Broad band antenna, 1948.

[13] A. P. King. Transmission, radiation and reception of electromagnetic waves, 1946.

[14] R. W. Masters. Antenna, 1947.

[15] W. Stohr. Broadband ellipsoidal dipole antenna, 1968.

[16] G. Robert-Pierre Marié. Wide band slot antenna, 1962.

[17] H. Schantz. *The Art and Science of Ultra Wideband Antennas.* Artech House, Norwood, MA, 2005.

[18] F. Lalezari. Broadband notch antenna, 1989.

[19] M. Thomas. Wideband arrayable planar radiator, 1994.

[20] H. F. Harmuth. Frequency independent shielded loop antenna, 1985.

[21] R. Garg, P. Bhartia, I. Bahl, and A. Ittipiboon. *Microstrip Antenna Design Handbook.* Artech House, 2001.

[22] P. J. Gibson. The Vivaldi aerial. In *Proceedings 9th European Microwave Conference*, pages 101–105, 1979.

[23] M. A. Peyrot-Solis, G. M. Galvan-Tejada, and H. Jardón-Aguiar. State of the art in ultra-wideband antennas. In *II International Conference on Electrical and Electronics Engineering (ICEEE)*, pages 101–105, 2005.

[24] E. Gazit. Improved design of the Vivaldi antenna. *IEE Proceedings*, 135(2):89–92, 1988.

[25] J. P. Weem, B. V. Notaros, and Z. Popovic. Broadband element array considerations for SKA. *Perspectives on Radio Astronomy Technologies for Large Antenna Arrays, Netherlands Foundation for Research in Astronomy*, pages 59–67, 1999.

[26] S. H. Choi, J. K. Park, S. K. Kim, and J. Y. Park. A new ultra-wideband antenna for UWB applications. *Microwave and Optical Letters*, 40(5):399–401, 2004.

[27] Y. Kim and D. H. Kwon. CPW-fed planar ultra wideband antenna having a frequency band notch function. *Electronics Letters*, 40(7):403–405, 2004.

[28] C. Ying and Y. P. Zhang. Integration of ultra-wideband slot antenna on LTCC substrate. *Electronics Letters*, 40(11):645–646, 2004.

[29] J. Yeo, Y. Lee, and R. Mittra. Wideband slot antennas for wireless communications. *IEE Proceedings Microwave and Antennas Propagation*, 151(4):351–355, 2004.

[30] J. Liang, C. C. Chiau, X. Chen, and C. G. Parini. Printed circular disc monopole antenna for ultra-wideband applications. *Electronics Letters*, 40(20):1246–1247, 2004.

[31] W. J. Lui, C. H. Cheng, Y. Cheng, and H. Zhu. Frequency notched ultra-wideband microstrip slot antenna with fractal tuning stub. *Electronics Letters*, 41(6):9–10, 2005.

[32] T. G. Ma and S. K. Jeng. Planar miniature tapered-slot-fed annular slot antennas for ultrawide-band radios. *IEEE Transactions on Antennas and Propagation*, 53(3):1194–1202, 2005.

[33] D. C. Chang, J. C. Liu, and M. Y. Liu. A novel tulip-shaped monopole antenna for UWB applications. *Microwave and Optical Technology Letters*, 48(2):307–312, 2006.

[34] S. Hu, C. L. Law, and W. Dou. A balloon-shaped monopole antenna for passive UWB-RFID tag applications. *IEEE Antennas and Wireless Propagation Letters*, 7:366–368, 2008.

[35] W.-J. Liu and Q.-X. Chu. A tapered CPW structure half cut disc UWB antenna for USB applications. In *Asia Pacific Microwave Conference*, pages 778–781. IEEE, 2009.

[36] T. Sugitani, S. Kubota, A. Toya, and T. Kikkawa. Compact planar UWB antenna array for breast cancer detection. In *2012 IEEE Antennas and Propagation Society International Symposium*, pages 1–2, 2012.

[37] S. Tripathi, S. Yadav, V. Vijay, A. Dixit, and A. Mohan. A novel multi band notched octagonal sshape fractal UWB antenna. In *2013 International Conference on Signal Processing and Communication*, pages 167–169, 2013.

[38] S. Y. Suh, W. L. Stutzman, and W. A. Davis. A new ultrawide-band printed monopole antenna: the planar inverted cone antenna (PICA). *IEEE Transactions on Antennas and Propagation*, 52(5):1361–1365, 2004.

[39] Z. N. Chen. Novel bi-arm rolled monopole for UWB applications. *IEEE Transactions on Antennas and Propagation*, 53(2):672–677, 2005.

[40] S. W. Su, K. L. Wong, and C. L. Tang. Ultra-wideband square planar monopole antenna for IEEE 802.16a operation in the 2–11 GHz band. *Microwave and Optical Technology Letters*, 42(6):463–465, 2004.

[41] F. W. Yao, S. S. Zhong, and X. X. L. Liang. Experimental study of ultra-broadband patch antenna using a wedge-shaped air substrate. *Microwave and Optical Technology Letters*, 48(2):218–220, 2006.

[42] N. Chahat, M. Zhadobov, R. Sauleau, and K. Ito. A compact UWB antenna for on-body applications. *IEEE Transactions on Antennas and Propagation*, 59(4):1123–1131, 2011.

[43] M. Botello-Perez, I. Garcia-Ruiz, and H. Jardón-Aguilar. Design and simulation of a 1 to 14 GHz broadband electromagnetic compatibility DRGH antenna. In *II International Conference on Electrical and Electronics Engineering (ICEEE)*, pages 118–121, 2005.

[44] M. Botello Pérez. Desarrollo de una antena de UWB para compatibilidad electromagnética y para el monitoreo del espectro radioeléctrico (in spanish). Master's thesis, Center for Research and Advanced Studies of IPN, Department of Electrical Engineering, Communications Section, 2005.

4

Developments in Ultra Wideband Antenna Theory

CONTENTS

4.1 Introduction

As has been shown in previous chapters, there are different approaches to designing UWB antennas. Particular types have been proposed for different applications, from wireless mobile communications and body area networks, to spectrum monitoring applications. Of all these UWB antennas, some have become more popular due to their performance and compact structure.

In order to discuss theoretical studies of UWB antennas, we must find a starting point. In classical narrowband antenna theory, the starting point is centered on the infinitesimal dipole, whose longitude $l << \lambda$ with λ the wavelength of the resonance frequency. From a Maxwell's equations solution for this radiating element, an analysis is derived to determine the parameters that describe it. In this case, the resonance frequency naturally becomes a central parameter. Nevertheless, this cannot apply in the case of UWB, because a wide bandwidth implies that there is more than one resonance frequency.

The most studied wideband antennas are the cylindrical dipole and the biconical antenna, as they present a simple geometry and offer a smooth transition between the transmission line and free space. Both antennas are closely related since, depending on the analysis carried out, one may replicate the other. For example, in order to increase the bandwidth of the cylindrical dipole, a uniform expansion of its conductor's diameters is applied, thus transforming it into a biconical antenna. On the other hand, when its aperture angle is $\alpha = 0$, it is transformed into a cylindrical dipole. So, some aspects relating to the biconical antenna in particular will be addressed, in order to present the background of theoretical concepts relating to UBW radiators.

Finally, another important area to consider in UWB theory relates to the time-domain response of UWB antennas. In fact, it is this topic that is commonly associated with UWB antenna theory, and many related works can be found in the open literature. Since both time-domain and frequency-domain responses impact upon the UWB pulse distortion, and consequently on the transmission rate of a UWB communication system, this topic will be dealt with in its own chapter (see Chapter 5), where more details, particularly those related to phase linearity, will be addressed in necessary breadth and depth.

4.2   UWB Bandwidth

Naturally, one of the most important parameters of UWB antennas is their bandwidth, because different parameters (radiation pattern, gain, impedance, polarization, etc.) can vary through the frequency. For example, Figure 4.1 shows the changes that the radiation pattern of a 1 GHz patch antenna experience at three different frequencies.

In general terms, an antenna is a resonant device, such that its impedance varies as a function of frequency, even though its feeder impedance remains inherently unchanged. As stated in Section 2.10, the interval of frequencies where the antenna is matched to its feeder is referred to as "impedance bandwidth." Usually, if an antenna presents 10% or fewer reflections with respect to its input signal, it is said that this antenna is well matched. Then, it is

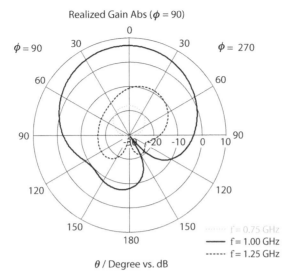

FIGURE 4.1
Variation of the radiation pattern of a 1 GHz patch antenna as a function of frequency.

fundamental to consider the $VSWR$ and reflection loss characterization through the interest frequency band, in order to determine the impedance matching. Both parameters depend on the the value of the reflection coefficient Γ, which was defined in Chapter 2.

Now, as was explained in Section 2.10, the classical bandwidth definition of any antenna is the difference between the upper and lower cut-off frequencies (f_H and f_L, respectively):

$$BW = f_H - f_L \qquad (4.1)$$

If the matching bandwidth is being considered, f_H and f_L can be taken as the -10 dB crossing points of $|\Gamma|$. It is worth remembering that it is common to represent the difference of frequencies given by (4.1) as the ratio $f_H : f_L$, which provides the number of times that f_H is higher than f_L (e.g. 2:1, 10:1, etc.). Of course, this BW ratio is not useful for defining the specific interval of frequencies over which an antenna is being designed, since a BW of 5:1, for instance, could represent a range from 200 MHz to 1 GHz, or equally from 2 GHz to 10 GHz.

On the other hand, from quality factor analysis for the UWB antenna, Schantz derived a relation between Q and BW as [1]

$$Q = \frac{f_0}{BW} \qquad (4.2)$$

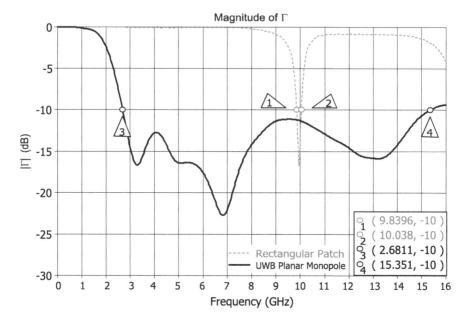

FIGURE 4.2
Coefficient reflection magnitude for a rectangular patch (narrowband) and for a UWB planar monopole.

where f_0 corresponds to the antenna resonance frequency (for UWB antennas, $f_0 = f_c$ with f_c, the central frequency). The expression (4.2) with $f_0 = f_c$ is compared to Q relations derived in [2–4] in Schantz's analysis of the Q-limit for UWB antennas. These comparisons serve to demonstrate the divergence between f_c, f_H and f_L in the UWB limit, and therefore that the determination of f_c in a UWB antenna can be complicated. Traditionally this frequency is simply obtained by means of the arithmetic average:

$$f_{c_\mu} = \frac{f_L + f_H}{2} \tag{4.3}$$

which is useful in the narrowband framework. Nevertheless, when the band is extended and the difference between f_L and f_H is one decade or more, the geometric average of f_c is more appropriate because the frequency span is given in a logarithmic scale. This average is then obtained by

$$f_{c_g} = \sqrt{f_L f_H} \tag{4.4}$$

which Schantz also demonstrates to correspond to the resonance frequency [1]. In order to illustrate the differences between (4.3) and (4.4), let us consider both a narrowband rectangular patch antenna and a UWB planar monopole,

TABLE 4.1

Comparison of central frequencies determined from Equations (4.3) and (4.4)

Antenna	f_L (GHz)	f_H (GHz)	f_{c_μ} (GHz)	f_{c_g} (GHz)	BW (GHz)
Rectangular patch	9.8396	10.038	9.9388	9.9383	0.1984
UWB planar monopole	2.6811	15.351	9.0160	6.4154	12.6699

whose reflection coefficient magnitude plots are shown in Figure 4.2. The frequencies f_L and f_H obtained from the -10 dB crossing points of these plots are presented in Table 4.1. In this table the values of BW, f_{c_μ} and f_{c_g} are determined from (4.1), (4.3), and (4.4), respectively.

As can be seen from the results for the narrowband antenna case, both the upper and the lower cut-off frequencies are very similar, thus presenting a narrow bandwidth. The similitude of these frequencies implies little difference between f_{c_μ} and f_{c_g}, such that any of them could be used in order to determine the central frequency. In contrast, for the UWB monopole, results are quite different. As can be seen, the bandwidth is one magnitude order larger than that corresponding to the narrowband antenna, which is produced by the larger separation between f_L and f_H. Naturally, this separation affects the results of the central frequency as determined by (4.3) or by (4.4), hence the values of f_{c_μ} and f_{c_g} present a higher variation (2.6006 GHz versus 500 kHz for the narrowband case).

Having addressed the importance of determining f_c as a geometric average, let us now consider the relationship between Q and BW as given in Equation (4.2). This equation expresses how the quality factor is reduced as the antenna bandwidth is increased. The physical interpretation of this inverse relation is that when Q is low (and therefore the bandwidth is larger), there is less reactive energy stored in the antenna and more radiated energy, as seen in Section 2.9.

However, it is worth noting that care must be taken when the value of Q is low, because, as Chu states in his work on the fundamental limits of antennas [2], although this factor shows the frequency dependence of the antenna (or a circuit), a low Q could not provide a precise interpretation. This fact was then addressed by Schantz [1], who presents some comments related to a low Q, particularly when $Q \to 1$ or less, in which case the UWB antenna performance must be evaluated based on the lower cut-off frequency.

Thus, in order to make sense of the Q factor concept for UWB antennas, two conditions should be considered [1]:

1. On one hand, the frequencies f_L and f_H are defined by the half-power or -3 dB points of the normalized impedance response.

2. On the other hand, the central frequency f_c must be determined as its geometric average, as given in (4.4).

Here it is important to present another concept related to the bandwidth, known as *fractional bandwidth* (*bw*). Strictly speaking, this term does not correspond to a bandwidth definition, but rather to a fraction of the total bandwidth with respect to its central frequency. In other words,

$$bw = \frac{BW}{f_c} = \frac{f_H - f_L}{\sqrt{f_L f_H}} \tag{4.5}$$

Finally, as was mentioned in Chapter 1, in order for a certain bandwidth to be considered in the UWB category, it should be $BW > 500\,\text{MHz}$ or $bw > 0.2$ as the FCC states [5].

4.3   Preliminary Concepts

In the late nineteenth century, it was found that revolution-surface radiator elements presented a wide bandwidth because of their high exclusion of reactive energy [1]. As was reviewed in Chapter 2, if the reactive energy is stored around an antenna, it is reflected, thus increasing the mismatching losses and also increasing the reactive part of the antenna impedance. In addition, as indirectly dictated by Equation (4.2), the larger the stored energy, the narrower the bandwidth. In other words, a low reactive energy means "better matching" and wider bandwidth [1]. This direct relationship between antenna efficiency in occupying the volume of the radiansphere, and its bandwidth, explains initially why an electric dipole can increase its bandwidth if its high/width ratio is closer to unity, increasing its volume in the radiansphere.

As was mentioned, the antenna bandwidth (which can be contained in a sphere of radius R) can be increased if the antenna, with its geometric structure, uses efficiently the most volume possible within the radiansphere [6]. This idea that a "fat" antenna has a wideband because it minimizes its stored energy comes from Harold Wheeler [7] who in the late 1950s identified the concept of *radiansphere*, which is a spherical frontier with a radius of $R = \lambda/2\pi$, whose center is a small dipole. This is exactly the radial distance where the radiated field component, or far field, and the near field, or reactive component, are equal in magnitude. Indeed, Wheeler stated: "Physically, it marks the transition between the "near field" inside and the "far field" outside" [8].

The concepts identified by Wheeler relating the radiation efficiency to the volume occupied by an antenna in the radiansphere, are based on two sentences that have been identified by Schantz as conceptually incorrect [1]:

1. "...Wheeler's thinking assumed that reactive energy is uniformly distributed inside the radian sphere."

2. "...Wheeler assumed that the only way to exclude reactive energy is to expand the antenna to occupy more volume."

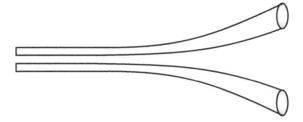

FIGURE 4.3
Representation of the transition of a transmission line to the biconical antenna.

These premises were the reason that radiators based on surface-revolution structures were initially considered during the initial search for wideband antennas, which also explains why the biconical antenna was one of the most studied antennas at that time. However, the Wheeler premises were later demonstrated to be limited, and most more recent UWB antenna designs only occupy a small fraction of the volume of the radiansphere.

During the years that Wheeler promoted his ideas of volumetric designs to increase the antenna bandwidth, an important principle was formulated based on the equivalence between the biconical and the dipole antennas as shown in Section 4.1: increasing the cross section of a dipole allows it to present a wider bandwidth. In general, there are three methods to deal with the finite thickness of a conductor, such that its radiation properties can be determined. The first method addresses it as a boundary problem, but this approach is useful only for radiators with ideal symmetric geometries, like a sphere, and cannot be used for relatively "more complex" geometries like cylinders, cones, etc. The second method, proposed by Schelkunoff [9], is based on a biconical antenna design, and cannot be applied to non-biconical geometries. With this method Schelkunoff represented an antenna as a transmission line made up of two conical-shape wires whose diameters are reducing, so forming a biconical antenna such as that shown in Figure 4.3. Thus, his solution is achieved through transmission line theory.

The third method does not present the limitations of the other methods. It is related to current distribution in a wire obtained from an integral equation, whereby the solution is reached through the Method of Moments (see Chapter 9), which is the basis of several softwares that simulate high frequency electromagnetic structures [6].

Years later, planar structures were considered to be a feasible solution to broadening the bandwidth of an antenna. Devices based on these structures can achieve the wide band aim if at least one of their dimensions is large [1]. Chen and Chia [10] summarize the techniques available to broaden the bandwidth of an antenna, grouping them into the following three approaches:

- Lowering the Q, by: modifying the radiator shape, thickening the substrate, reducing the dielectric constant, or increasing the losses.

- Using impedance matching, by: inserting a matching network, adding tuning elements, or using slotting and notching patches.

- Introducing multiple resonances, by: using parasitic elements, slotting patches, impedance networks, or using an aperture, proximity coupling.

Up to this point, the biconical antenna has been mentioned as one of the structures that allows wide bandwidths to be achieved. Therefore, some of its most relevant aspects are briefly presented below.

4.4   Biconical Antenna

As previously mentioned, the biconical antenna is the classic radiating device through which theoretical analyses have been developed in the field of wideband antennas [9]. Hence, before addressing the corresponding theory of UWB radiators, let us introduce some details associated with the biconical antenna.

The analysis of this antenna is based on transmission line theory, due to the fact that a biconical antenna can be seen as a uniformly expanded transmission line, as was explained in Sections 4.1 and 4.2. Now, as is well known, the current is distributed along the cones' surfaces, and the analysis of radiated fields assumes a transverse electromagnetic (TEM) excitation of the wave. Then, from the analysis of voltage produced at a distance r from the origin (taking the origin as the union point of the two cones), and the current over the cones' surface, it is possible to derive the characteristic impedance as [6]:

$$Z_c = \frac{V(r)}{I(r)} = \frac{\eta}{\pi} \ln \left[\cot \left(\frac{\alpha}{4} \right) \right] \qquad (4.6)$$

where η is the free space impedance. As can be seen from Equation (4.6), the characteristic impedance is independent of the radial distance r, and thus it can also represent the characteristic impedance of a biconical antenna of infinite structure. However, Equation (4.6) shows dependence on antenna geometry (through the aperture angle α), and we must remember that the derivation of these equations is based on transmission line theory and the assumption of TEM field. Thus, many early theoretical approaches for distinct designs like the triangular, bow-tie, and cylindrical dipole were based on the biconical theory.

Nevertheless, all these conditions do not necessarily correspond to some recent proposals of UWB antennas presented in Chapter 3. As stated in that chapter, many UWB radiator designs are based on planar or planarized structures which, fortunately, solve the need for compactness in modern applications.

4.5 Planar Monopole Structure as a Basic Element of UWB Antenna Theory

In recent decades, once different authors pointed out that a "fat" antenna is not the only possible way of achieving wide bandwidths, planar structures became very popular UWB radiators. The rectangular planar monopole antenna, whose geometry is depicted in Figure 4.4, is taken as the basis of the theoretical work presented in this chapter. This antenna was originally proposed by Dubost and Zisler in 1976 [11], and since then it has been widely studied by the scientific community. It has been found that it can be theoretically explained by means of the following two approaches:

- A planar monopole antenna can be seen as a microstrip antenna, which has a very thick air substrate. The shape and size of this antenna (and other designs derived from it) can be adjusted in order to achieve a certain impedance matching.

- A planar monopole antenna can be compared to a cylindrical one with a very large effective diameter, and to which the solid-planar correspondence

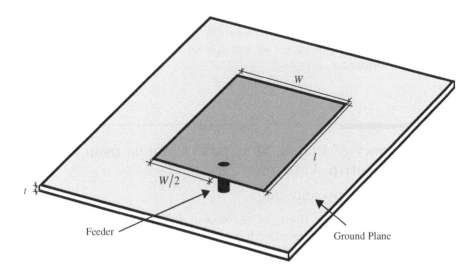

FIGURE 4.4
Geometry of a rectangular planar monopole.

Solid - Planar

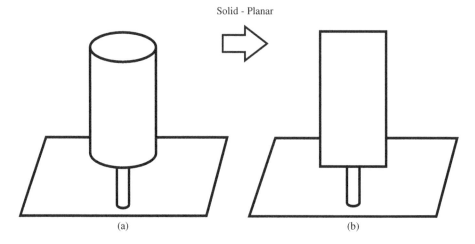

(a)   (b)

FIGURE 4.5
Example of the solid-planar principle. (a) Volumetric structure, (b) planar structure.

principle has been applied. A traditional monopole antenna is usually made up of a thin wire mounted over a ground plane, and whose bandwidth is increased as its diameter is extended.

The *solid-planar correspondence principle*, or *solid-planar principle*, mentioned in the last paragraph, states that for any surface-revolution structure, there is an equivalent planar antenna [1]. In other words, it is possible to achieve a a volumetric radiator structure equivalent to its corresponding planar version. Figure 4.5 shows an example of two monopole antennas with similar performances.

4.6   Theory of Planar Monopole Antenna from a Microstrip Antenna

4.6.1   Microstrip antenna

In order to develop the theory of the planar monopole antenna, let us first state some concepts associated with microstrip antennas. These concepts and equations can be found in diverse materials (e.g. [6, 12]).

As was seen in Chapter 2, the basic configuration of a microstrip antenna consists of a metallic, electrically thin patch printed over a dielectric substrate. This patch is the radiator element of the antenna, whose ground plane is under the substrate. The substrate thickness is always less than the wavelength of the signal. Usually, the main dimensions of the patch are equal to one half of a wavelength at the frequency of operation. The antenna characteristics depend on the excited operation mode, the dimensions and shape of the radiator, the substrate thickness and dielectric constant, and the feed method.

Microstrip antenna theory is based on the *transmission line model* (TLM) or the *cavity model* (CV) [13]. The equations in this chapter follow the TLM, in which the antenna is considered fundamentally as an open section of a transmission line with a longitude l, width W and a substrate thickness t. Thus, the resonance frequency in this model is obtained through its transverse magnetic mode (TM_{m0}) [1] from the following expression [14]:

$$f_{rm} = \frac{mc}{2(l + \Delta l)\sqrt{\varepsilon_{eff}}} \qquad (4.7)$$

where c is the light speed, m is an integer number different from zero, l is the longitude of the patch antenna, Δl is the equivalent length after considering the outline fields at the open ends of the TLM, and ε_{eff} is the effective relative dielectric permittivity, which is given by

$$\varepsilon_{eff} = \frac{\varepsilon_r + 1}{2} + \frac{\varepsilon_r - 1}{2}\left(1 + 10\frac{t}{W}\right)^{-\gamma\sigma} \qquad (4.8)$$

where ε_r is the relative permittivity of the substrate. The variables that involve the exponent $\gamma\sigma$ are determined by the following expressions:

$$\gamma = 1 + \frac{1}{49}\log\left\{\frac{\left(\frac{W}{t}\right)^4 + \left(\frac{1}{52}\frac{W}{t}\right)^2}{\left(\frac{W}{t}\right)^4 + 0.432}\right\} + \frac{1}{18.7}\log\left\{1 + \left(\frac{1}{18.1}\frac{W}{t}\right)^3\right\} \qquad (4.9)$$

$$\sigma = 0.564\left(\frac{\varepsilon_r - 0.9}{\varepsilon_r + 3}\right)^{0.053} \qquad (4.10)$$

Finally, the expression that relates Δl is given by

$$\frac{\Delta l}{t} = 0.412\frac{(\varepsilon_{eff} + 0.3)\left(\frac{W}{t} + 0.264\right)}{(\varepsilon_{eff} - 0.258)\left(\frac{W}{t} + 0.8\right)} \qquad (4.11)$$

Basically, this patch antenna is a narrowband device to which an impedance matching technique has been applied in order to increase its bandwidth. Essentially, this bandwidth increase can be attributed to a low Q together with multiple excited resonance modes. One way of reducing the quality factor of a patch microstrip antenna is by inserting a wideband matching network between the antenna and its feeder. On the other hand, if two or more

---

[1]It is important to point out that the dominant mode of microstrip antennas is the TM_{10}.

adjacent resonance modes are simultaneously and effectively well excited, the resulting bandwidth could be more than double the impedance bandwidth achieved using only one resonance.

Matching techniques commonly used to decrease the value of Q in a microstrip are: modifying the radiator shape, increasing the substrate thickness, and decreasing the dielectric constant or allowing more losses. It is worth noting that there are also other impedance matching techniques such as adding tune elements, or using slots and notches on the patch antenna. Some of these are covered below.

Research has shown that the radiator shape affects the impedance bandwidth of the antenna, even for identical dimensions [12,13]. However, varying the radiator shape is a limited technique for the purpose of increasing bandwidth, since it critically affects radiation characteristics, and so is in fact rarely used in practice.

As regards the possibility of inserting a matching network without affecting the radiator shape, either by slots or by notching, this method has the advantage of a simplified antenna design (note that the radiator shape is not modified). The disadvantages are, however, that the total size is increased, and the efficiency is reduced due to the increase in losses.

Other research has shown impedance bandwidth increasing through substrate thickness enlargement. Nevertheless, the increasing bandwidth is inverted once the substrate thickness exceeds a certain threshold [13], hence this option cannot be used for any impedance bandwidth. For example, in a perpendicular fed microstrip patch antenna, the enlargement of the substrate thickness makes the connector seem larger, and therefore a higher inductance is reflected at the input, limiting a possible adequate impedance matching.

4.6.2   Planar monopole antenna

Since wideband planar monopoles can be considered as a modified microstrip antenna, let us now analyze how microstrip antenna theory can be adapted to a planar monopole (remember that a microstrip antenna is in principle a resonant antenna). As was emphasized above, the enlargement of the substrate thickness of microstrip antennas produces a wider bandwidth. Hence, if a rectangular planarized radiator is fed by a coaxial line with a perpendicular ground plane, the substrate (in this case air) would have a large thickness, and an effective dielectric constant equal to the unity.

In the case of conventional microstrip antennas, when the substrate thickness, t, is increased (see Figure 4.6a), the height of the inner conductor of the coaxial line is also increased. This longer inner conductor produces a higher inductance, making it difficult to achieve adequate impedance matching. Naturally, this highly inductive input impedance can be eliminated if the patch antenna is fed with a coaxial line whose inner conductor longitude, h, is shorter. In this case, the antenna would be fed through the side, as shown in Figure 4.6b, thus making an additional perpendicular ground plane necessary.

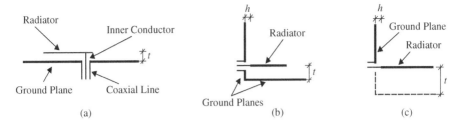

FIGURE 4.6
Geometry of the microstrip antenna: (a) Air substrate, (b) lateral feeding and perpendicular ground planes, (c) planar monopole antenna.

If the distance t is too large, then the effects of the original ground plane could be ignored, and it could be removed, achieving a configuration similar to that of a rectangular planar monopole antenna (see Figure 4.6c).

If the planar monopole antenna is seen as a microstrip antenna with air substrate and an orthogonal ground plane, the analytical methods valid for microstrip antennas can be applied with some modifications. The equation that determines the resonance frequency of a rectangular microstrip antenna can be applied to that shown in Figure 4.6c, for instance. Thus, the theoretical resonance frequency (in GHz) for the fundamental propagation mode can be calculated based on the following expression, with l_e being the effective longitude in cm [12]:

$$f = \frac{30}{2l_e\sqrt{\varepsilon_{eff}}} \qquad (4.12)$$

For the two orthogonal ground planes, $l_e = l + \Delta l + h$ where Δl is associated with the outline fields present only on one side, whereas on the other side it is restricted to h (due to the orthogonal ground plane). For wide patch radiators ($W/t > 10$) with $\varepsilon_r = 1$, $\Delta l \approx t$. Then, the larger the t, the larger the bandwidth showing behavior similar to a planar monopole antenna.

4.7 Planar Monopole Antenna from a Cylindrical Monopole Antenna

4.7.1 Resonance frequency

As has been shown, the operation frequency is one of the most important parameters in the design of any antenna, and it can be evaluated in the dominant resonance frequency. For example, the operation frequency (resonance frequency) of a narrowband thin wire monopole can be determined from [10]:

$$f = \frac{c}{4l} \qquad (4.13)$$

where l is the longitude of the radiator. Therefore, the operation frequency and the bandwidth obtained from this central frequency are enough to start the design of this antenna.

In contrast, in the case of UWB planar monopoles, it is necessary to evaluate the lower cut-off frequency of the impedance bandwidth, which could be difficult to determine due to the multiple resonances that a UWB antenna presents. Thus, taking into consideration the monopole width, the dominant frequency cannot be determined with the Equation (4.13), since it is only an approximation, useful for a thin monopole with a circular cross section. For wideband planar monopoles, the evaluation of the dominant frequency becomes complicated because it depends on many factors, such as the shape and dimensions of the radiator, among others.

Now, as stated in Section 4.5, a planar monopole antenna can be compared to a cylindrical monopole (the classical wire monopole) if the solid-planar principle is applied such that its diameter is very large. Hence, in order to determine the lower cut-off frequency of a planar structure, the analysis presented here will take the cylindrical monopole as the starting point, which is a widely-studied antenna.

4.7.2 Lower cut-off frequency for different planar antenna shapes

It is known that a square planar monopole presents a lesser bandwidth than that corresponding to a circular monopole. Nevertheless, the radiation pattern of the former suffers less degradation within its bandwidth [15]. In addition, the solid-planar principle explained in Section 4.5 can be applied to a cylindrical monopole with a very large effective radius, r_d, in order to obtain a square planar monopole. In this way, several theoretical expressions used for a classical wire monopole can be extrapolated to a planar square antenna, if the radius of the former is extended to form a wide cylinder, and then the solid-planar principle applied to it. Thus, the approximate lower cut-off frequency for some simple geometry planar monopoles can be determined from the expression of the monopole longitude for a real input impedance [1]:

$$l = 0.24\lambda F \qquad (4.14)$$

where F is a term known as the *longitude-radius equivalence factor*, which varies from 0.86 for a square planar monopole, up to 0.99 for a thin wire monopole and is given by [16]:

$$F = \frac{l}{r_d + l} \qquad (4.15)$$

So, the design procedure consists of making equal the areas of the planar monopole antenna geometry in question, and the cylindrical wire with longitude l (the area taken into account for the cylindrical wire is that of its lateral

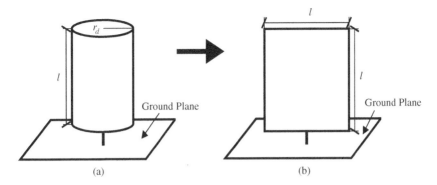

FIGURE 4.7
Representation of the equalizing of areas for (a) Basic cylindrical monopole with very large effective radius and (b) square planar monopole.

side, $2\pi r_d\, l$, provided that we are transforming this volumetric structure to a flat one). After that, the radius r_d is derived and it is substituted into Equation (4.15). Finally, this value of F is substituted in (4.14), with which it is possible to calculate the lower cut-off frequency, f_L.

4.7.2.1 Example 1

Let us consider the square planar monopole with dimensions $l \times l$ shown in Figure 4.7b. By making equal its area and that corresponding to the basic cylindrical wire monopole, i.e., $l^2 = 2\pi r_d\, l$, it is easy to obtain r_d:

$$r_d = \frac{l}{2\pi} \tag{4.16}$$

and substituting (4.16) into (4.15) it is directly obtained that $F = 0.8626$. Now, by applying this value to (4.14), remembering that $\lambda = c/f_L$, and if l is expressed in mm and f_L in GHz, it is found that

$$f_L = \frac{62.11}{l} \tag{4.17}$$

This equation was presented by Ammann in [16] for a square planar monopole with a very large ground plane.

4.7.2.2 Example 2

Let us now take an ellipsoidal planar monopole with semi-major axis a_1 and semi-minor axis b_1, like those presented in [17, 18], and whose geometry is depicted in Figure 4.8 for three possible cases. By making its area equal to that of the cylindrical wide monopole, we can easily derive that

$$r_d = \frac{a_1 b_1}{2l} \tag{4.18}$$

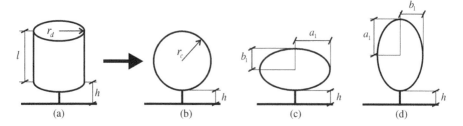

FIGURE 4.8
Representation of the equalization of areas for (a) Basic cylindrical monopole with very large effective radius, and elliptic monopoles with semi-major axis a_1 and semi-minor axis b_1; (b) $r_c = a_1 = b_1$; (c) a_1 on the horizontal axis; (d) a_1 on the vertical axis.

From which substituted into (4.15), the factor F is given by

$$F = \frac{2l^2}{a_1 b_1 + 2l^2} \tag{4.19}$$

and by substituting (4.19) into (4.14) and following the same steps and considerations of Example 1 above (all dimensions in mm and f_L in GHz), it is derived that

$$f_L = \frac{0.48l}{2l^2 + a_1 b_1} \tag{4.20}$$

Equation (4.20) was not presented in [18]; instead, the relationship between f_L and l suggested there, is given by

$$f_L = \frac{72}{r_d + l} \tag{4.21}$$

In principle, Equation (4.21) contains r_d, which could be substituted by the relation with a_1 and b_1, as expressed in (4.18). In this case, Equations (4.20) and (4.21) are equivalent except for the constant values of 0.48 and 72, respectively.

4.7.2.3   Example 3

The following example is a trapezoidal radiator that comes from the cylindrical monopole, and over which the solid-planar principle has been applied as shown in Figure 4.9. Thus, by equalizing their areas, i.e.,

$$\left(\frac{W_1 + W_2}{2} \right) l = 2\pi r_d l \tag{4.22}$$

then,

$$r_d = \frac{W_1 + W_2}{4\pi} \tag{4.23}$$

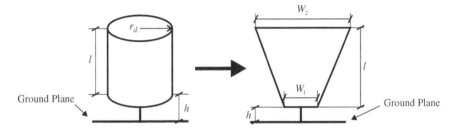

FIGURE 4.9
Representation of the equalization of areas for (a) Basic cylindrical monopole
with a very large effective radius and (b) trapezoidal planar monopole.

and by substituting (4.23) in (4.15), it is found that

$$F = \frac{4\pi l}{W_1 + W_2 + 4\pi l} \tag{4.24}$$

If (4.24) is substituted into (4.14), and following again the same steps and
considerations of Example 1 above (all dimensions in mm and f_L in GHz),
the following expression is easily obtained:

$$f_L = \frac{904.77}{W_1 + W_2 + 4\pi l} \tag{4.25}$$

which corresponds to the design equation for an l height trapezoidal monopole
presented in [19].

4.7.2.4 Example 4

The final example consists of applying the solid-planar principle to the cylindrical monopole in order to make up a $W \times l$ rectangular radiator as depicted
in Figure 4.10. By making their areas equal, it is easy to derive that

$$r_d = \frac{W}{2\pi} \tag{4.26}$$

which substituted in (4.15) results as

$$F = \frac{2\pi l}{W + 2\pi l} \tag{4.27}$$

and therefore by substituting (4.27) into (4.14), and making the same assump-
tions as in Example 1, the following expression is directly determined:

$$f_L = \frac{144\pi}{W + 2\pi l} \tag{4.28}$$

where dimensions W and l are in mm and f_L in GHz.

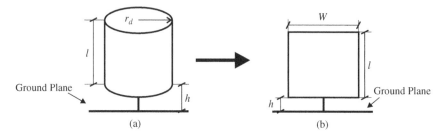

FIGURE 4.10
Representation of the equalization of areas for (a) Basic cylindrical monopole
with a very large effective radius and (b) rectangular planar monopole.

4.8   Some Factors That Influence UWB Antenna Performance

Now that we have discussed different concepts associated with UWB theory,
let us address the influence of different factors on the performance of antennas.

4.8.1   Influence of the radiator

The distance from the radiator feeder to the ground plane h was studied
by Ammann [16], who explains that it impacts the bandwidth of a square
monopole. Moreover at a distance of $h = 25\,\text{mm}$, it is reported that there
is an optimal bandwidth. Ammann also studied the effects of h on f_L and
f_H, and found that whereas f_L is independent of h, f_H does present a strong
dependence on this distance.

Another interesting piece of research focused on the effect of the tilt angle β
of the radiator (see Figure 4.11) on the input impedance. According to results
presented by Chen [20] for planar monopoles of different shapes (square, disc,
diamond, triangle and bow-tie-like), as this angle is reduced, the frequency f_L
is shifted towards a lower band, causing the bandwidth to increase. Table 4.2
shows the frequency range obtained for different angles β.

As regards the frequency f_H, it is known that this parameter depends on
the geometry of the planar element near to the ground plane and the radiator
feeder, where there is a high density of currents [16, 21]. For this reason, this
property has been exploited in the research field. An example of the above is
the beveling in a symmetric or asymmetric form on the bottom of the radiator
of a square or rectangular planar monopole (as shown Figure 4.12), where the
angle ψ defines the "grade" of the beveling on the radiator (hence it is called
the *Beveling Angle*) [22].

By using the technique reported in [21], different bevelings were applied
to a square monopole on both sides of the antenna in order to evaluate its

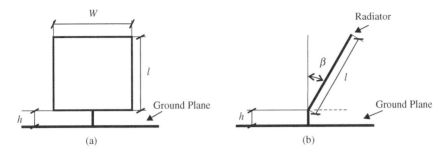

FIGURE 4.11
Square planar monopole with a tilt angle β on its radiator: (a) Frontal view,
(b) lateral view.

TABLE 4.2
Bandwidth variation as a function of β for a square
planar monopole

Tilt angle β (degrees)	Frequency range[†] (GHz)	BW (GHz)
0	1.50 − 3.10	1.60
22.5	1.52 − 3.10	1.58
45.0	2.25 − 3.10	0.85
67.5	2.60 − 3.25	0.65
90.0	3.25 − 3.50	0.25

[†]Approximated values taken from the measured $VSWR$ reported in [20]

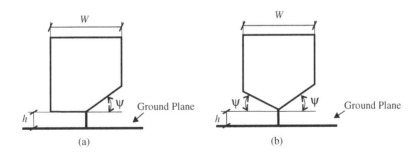

FIGURE 4.12
Beveling technique in a square planar monopole: (a) Asymmetric bevel, (b)
symmetric bevel.

TABLE 4.3
Measured bandwidth for different angles ψ for a square planar
monopole [21]

Bevel angle ψ (degrees)	Frequency range for asymmetrical bevel (GHz)	BW (GHz)	Frequency range for symmetrical bevel (GHz)	BW (GHz)
0	$2.35 - 4.95$	2.60	$2.35 - 4.95$	2.60
10	$2.20 - 5.30$	3.10	$2.12 - 5.95$	3.83
20	$2.19 - 5.75$	3.56	$2.11 - 6.75$	4.64
30	$2.17 - 5.97$	3.80	$2.10 - 7.25$	5.15
40	$2.17 - 6.00$	3.83	$2.10 - 12.50$	10.4

impact on the bandwidth. Table 4.3 shows the obtained results, where the
dependence on f_H as a function of ψ is clear: the higher the angle ψ, the
higher the frequency f_H.

4.8.2   Ground plane

Up to here, the term *radiator* has been used indistinctly to make reference to
an antenna. Indeed, it is so; but in the case of any monopole made up of a
radiator and its ground plane, strictly speaking, both of them form an antenna.
This fact is important to mention here, because the currents on the ground
plane affect the antenna performance; thus, this component contributes to
the whole radiation process, and it should be considered as determining the
electrical size of the antenna.

Some interesting research presented in 2009 by Best [23] shows some results of the effects of the ground plane and antenna position on the antenna
performance for single wire, planar and planarized structures used on classical
wireless communications bands. Although the dimensions of the different configurations considered in his work are not necessarily optimized for a specific
application, his results show some important behaviors relevant here:

- The size of the ground plane for wire antennas does not present a significant
  effect on the input impedance and bandwidth, but it does on the radiation
  characteristics. For instance, a reduction in the size of the ground plane
  degrades the omnidirectional feature.

- For device-integrated antennas, like PIFA (planar inverted F antenna), the
  ground plane size does have an impact on the antenna bandwidth, provided
  that the lower cut-off frequency is dependent on this dimension. This is in
  accordance with the discussion presented in Section 4.7.

- Since the main contribution of the currents is concentrated near the radiator,
  if it is located on a corner of the ground plane, standing-waves are formed
  on the antenna edges close to it, which cause an alteration of the radiation

pattern. Small frequency shifts are also presented due to this position change, which naturally affects the antenna bandwidth.

- In the case of planarized antennas, a small variation in the lower cut-off frequency could be presented as a function of the ground plane size.

- Naturally, the feeder type also affects the performance of planarized antennas, provided that the current's distribution is altered depending on the feeder features.

Bibliography

[1] H. Schantz. *The Art and Science of Ultra Wideband Antennas.* Artech House, Norwood, MA, 2005.

[2] L. J. Chu. Physical limitations of omni-directional antennas. *Journal of Applied Physics*, 10:1163–1175, 1948.

[3] R. F. Harrington. Effect of antenna size on gain, bandwidth, and efficiency. *Journal Res. NBS*, 64D:1–12, 1960.

[4] J. S. McLean. A re-examination of the fundamental limits on the radiation q of electrically small antennas. *IEEE Transactions on Antennas and Propagation*, 44(5):672–676, 1996.

[5] FCC. US 47 CFR part 15 subpart F §15.503d ultra-wideband operation. Technical report, Federal Communications Commission, 2003.

[6] C. A. Balanis. *Antenna Theory: Analysis and Design.* John Wiley & Sons, 3rd edition, 2005.

[7] H. A. Wheeler. Fundamental limitations of small antennas. *Proceedings of the IRE*, 35(12):1479–1484, 1947.

[8] H. A. Wheeler. The radiansphere around a small antenna. *Proceedings of the IRE*, 47(8):1325–1331, 1959.

[9] S. A. Schelkunoff. *Electromagnetic Waves.* D. van Nostrand Company, Inc., 1943.

[10] Z. N. Chen and M. Y. W. Chia. *Broadband Planar Antennas: Design and Applications.* Jonh Wiley & Sons, Sussex, England, 2006.

[11] G. Dubost and S. Zisler. *Antennas: A Large Band.* Masson, New York, 1976.

[12] R. Garg, P. Bhartia, I. Bahl, and A. Ittipiboon. *Microstrip Antenna Design Handbook.* Artech House, 2001.

[13] D. M. Pozar. *Microstrip Antennas: Analysis and Design.* IEEE-John Wiley & Sons, New York, 1995.

[14] G. Ramesh, B. Prakash, B. Inder, and I. Apisak. *Microstrip Antenna Design Handbook.* Artech House, 2001.

[15] M. Hammoud, P. Poey, and F. Colomel. Matching the input impedance of a broadband disc monopole. *Electronics Letters*, 29:406–407, 1993.

[16] M. J. Ammann. Square planar monopole anetnna. In *IEE National Conference on Antennas and Propagation*, pages 37–38, 1999.

[17] T. Y. Shih, C. L. Li, and C. S. Lai. Design of an UWB fully planar quasi-elliptic monopole antenna. In *Proc. Int. Conf. Electromagnetic Applications and Compatibility*, 2004.

[18] N. P. Agrawall, G. Kumar, and K. P. Ray. Wide-band planar monopole antennas. *IEEE Transactions on Antennas and Propagation*, 46(2):294–295, 1998.

[19] J. A. Evans and M. J. Ammann. Planar trapezoidal and pentagonal monopoles with impedance bandwidths in excess of 10:1. In *IEEE International Antennas and Propagation Symposium*, volume 3, pages 1558–1561, 1999.

[20] Z. N. Chen. Experiments on input impedance of tilted planar monopole antenna. *Microwave and Optical Technology Letters*, 26(3):202–204, 2000.

[21] M. J. Ammann. Control of impedance bandwidth of wideband planar monopole antennas using a beveling technique. *Microwave and Optical Technology Letters*, 30(4):229–232, 2001.

[22] M. A. Peyrot-Solis. *Investigación y Desarrollo de Antenas de Banda Ultra Ancha (in Spanish).* PhD thesis, Center for Research and Advanced Studies of IPN, Department of Electrical Engineering, Communications Section, Mexico, 2009.

[23] S. R. Best. The significance of ground-plane size and antenna location in establishing the performance of ground-plane-dependent antennas. *IEEE Antennas and Propagation Magazine*, 51(6):29–43, 2009.

5

Phase Linearity

CONTENTS

5.1   Time Domain and Frequency Domain

In the context of narrowband, concepts of antennas are usually only defined in the frequency domain because their parameters present little variation through the operational band of interest. As an example, let us recall the discussion of the central frequency in Section 4.2, where it was found that both the upper and the lower cut-off frequencies are very close for a rectangular patch antenna. Nevertheless, when the antenna bandwidth is broadened, some antenna responses could significantly vary (see Figure 4.1 of Chapter 4, for instance). Thus, any variation in the frequency of the signal has an impact on it in time. Hence it is important to analyze the antenna response not only in the frequency domain, but also in the time domain.

5.1.1   The Fourier transform

The Fourier analysis is one of the pillars supporting many areas related to signal processing, such as telecommunications, radar, imaging and so on, since it provides the means to understand the behavior of any analogue or digital signal both by its duration and shape (time domain) and by its spectral response (frequency domain). As is widely known, there is a duality between both domains, which can be determined through the Fourier transform. Thus, given a certain signal defined in the time domain $f(t)$, its spectral density $F(j\omega)$, with ω the angular frequency, is determined by the Fourier integral transformation:

$$F(j\omega) = \int_{-\infty}^{+\infty} f(t)\, e^{-j\omega t} dt \tag{5.1}$$

On the other hand, if the spectral density $F(j\omega)$ is given, it is possible to obtain its signal in the time domain by means of the inverse Fourier transform:

$$f(t) = \frac{1}{2\pi} \int_{-\infty}^{+\infty} F(j\omega)\, e^{j\omega t} d\omega \tag{5.2}$$

and it is a common practice to represent this transformation pair as $f(t) \longleftrightarrow F(j\omega)$. There is a wide selection of excellent references in this field in the open literature (e.g., [1–4]), and it is not the aim of this book to present an extended discussion of the issue of signal theory. Here instead, some classical concepts are noted in order to support the explanation of the relationship between a UWB antenna and the pulse distortion that it introduces.

5.1.2   Short time duration, wide spectrum

An important concept to explain is the scaling property. This property indicates that if a function in the time domain $f(t)$ is scaled by a factor a, i.e., $f(at)$, and if $f(t) \longleftrightarrow F(j\omega)$, then

$$f(at) \longleftrightarrow \frac{1}{|a|} F\left(\frac{j\omega}{a}\right) \tag{5.3}$$

which states that the effect of compressing a signal in one domain, is to expand it in the other domain. In other words, the compression of a function in the time domain by a certain value is equivalent to this function varying more rapidly at a rate related to the same factor, and hence the frequency components are proportionally increased.

Figures 5.1 and 5.2 show this property applied to a rectangular pulse when its width or duration (in time units) is $\tau = 1$ and $\tau = 0.3$, respectively. As can be seen in these figures, when the pulse width is reduced (i.e., it is faster), the corresponding spectral content is affected both in amplitude and the first zero crossing. In terms of the latter, a wider main spectral pulse is observed when

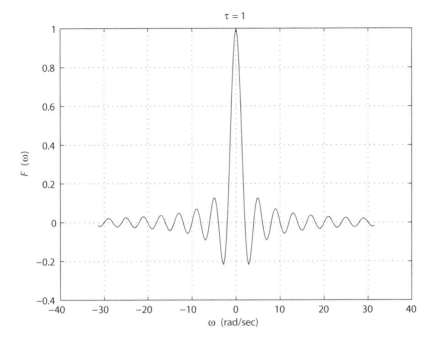

FIGURE 5.1
Spectrum of a rectangular pulse with a duration of $\tau = 1$ time units.

$\tau = 0.3$ in comparison with the case of $\tau = 1$, as was expected according to the scaling property of the Fourier transform.

Therefore, if the bandwidth is extended by several MHz (more than 500 MHz for UWB applications), short pulses are needed. In a variety of works it has been explained that pulses with a duration in the order of a few nanoseconds should be considered for ultra wideband systems (the FCC states a duration of between $0.1 - 2\,\mathrm{ns}$ [5]), for which reason they are commonly referred to as impulses or *ultra-short pulses*.

5.1.3 Impulse response and transfer function

The behavior of a system is represented by its impulse response (time domain), or by its transfer function (frequency domain), when a signal travels through it and whose analysis is based on linear or quasi-linear systems theories. In the case at hand, the antenna corresponds to the system that could introduce some effects on the signal to be transmitted, and consequently we can state that an antenna possesses an impulse response $h(t)$, and a transfer function $H(j\omega)$ (consider Figure 5.3 as a graphic description of the above). In fact, in narrowband antennas, this type of analysis is usually neglected provided that

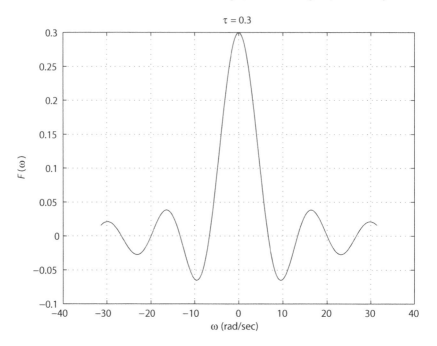

FIGURE 5.2
Spectrum of a rectangular pulse with a duration of $\tau = 0.3$ time units.

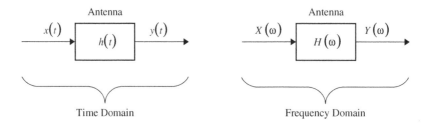

FIGURE 5.3
Impulse response and transfer function of an antenna.

the features of them do not present a dependence on the frequency. Instead, as different authors have pointed out ([6, 7]), the transient state cannot be ignored in a UWB system, and so the impulse response of UWB antennas is addressed in a range of work. This fact is attributed to the short duration of the pulses radiated by this type of antenna, which could severely affect some of the characteristics of the generated pulse, and consequently it is important to analyze both responses.

5.2 Characteristics of the Impulse Response of a UWB Antenna

As was noted in Section 5.1.2, pulses used in UWB applications have a short duration. It has been noted, then, that the transient response cannot be neglected in these applications. Therefore, some important quantities associated with the impulse response $h(t)$ of UWB antennas are presented here, based on [7] and summarized as follows:

- Peak value of the envelope: This parameter is simply the maximum value of the strongest peak of $h(t)$. It is desirable that the envelope of $h(t)$ is as high as possible.

- Envelope width: This quantity provides a measure of the broadening of the radiated impulse, and is defined as the width of the magnitude of the envelope at half maximum (τ_{FWHM}). Values of τ_{FWHM} not exceeding a few hundred picoseconds are suggested.

- Ringing: Term related to the oscillations of the radiated pulse after the main peak (see Figure 5.4). It is highly desirable that the ringing value is as small as possible (less than a few τ_{FWHM}), because this quantity represents a waste of non-transmitted energy.

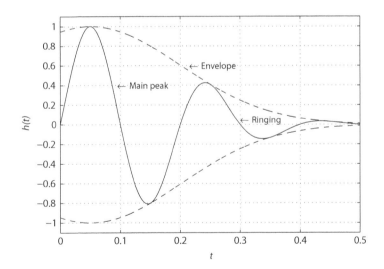

FIGURE 5.4
Ringing on an impulse response.

Ideally, a delta function would be desirable for $h(t)$, in such a way that the antenna output signal is the same as the signal supplied to it (time convolution of a delta function with other signal produces the same signal). Naturally, a delta function does not exist in practice, so an antenna always introduces a certain distortion to the signal that travels through it, as will be explained next in Section 5.3.

5.3   Pulse Distortion

Before discussing this topic, let us briefly comment on UWB pulses. There are different proposals of UWB pulses that could be implemented in practice, with the Rayleigh (differentiated Gaussian) pulse being one of the most discussed. This pulse is given by the following expression [8]:

$$v_n(t) = \frac{d^n}{dt^n}\left[e^{-\left(\frac{t}{\sigma}\right)^2}\right] \tag{5.4}$$

For $n = 0$, corresponding to a Gaussian pulse,

$$v_0(t) = e^{-\left(\frac{t}{\sigma}\right)^2} \tag{5.5}$$

In (5.4) σ stands for the time when the Gaussian pulse $v_0(t) = e^{-1}$. The first-order Rayleigh pulse is given when $n = 1$, so

$$
\begin{aligned}
v_1(t) &= -\frac{2}{\sigma^2}t\, e^{-\left(\frac{t}{\sigma}\right)^2} \\
&= -\frac{2}{\sigma^2}t\, v_0(t)
\end{aligned}
\tag{5.6}
$$

Figure 5.5 depicts the curves corresponding to the Gaussian (for $\sigma = 50\,\text{ps}$) and the Rayleigh ($\sigma = 110\,\text{ps}$) pulses. As indicated in this figure, normalized amplitude values are being plotted, which is done in order to compare them on an equal basis. As can be seen in Figure 5.5, these pulses have a short duration, and are made up of both a positive half cycle and a negative half cycle (the reason why they are known as *monocycles*). This figure also shows that these pulses fall to zero out of the monocycle duration (this characteristic is clearer for the Gaussian pulse). An important point here is that they must comply with the UWB spectral emissions masks stated by the FCC [5]. In order to evaluate this fundamental aspect, it is necessary to obtain the Fourier transform of (5.5) and (5.6), or whichever other pulse. Chen et al. showed that pulses generated from (5.4) (with certain limits on the value of σ) present adequate spectral characteristics within the FCC masks [8].

Although it is not the objective here to present all the approaches reported in the open literature relating to this issue, a wide selection of references are recommended for the interested reader: [6, 8–14], to mention just a few.

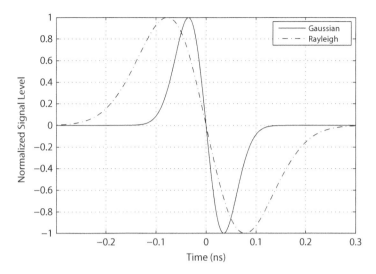

FIGURE 5.5
Time domain waveforms of Gaussian and Rayleigh pulses.

The problem of pulse distortion has been a subject of interest for several authors, who have reported their results in different forums [7,15–20]. In order to describe how a pulse can be distorted, let us first explain the pulse dispersion suffered by a pulse transmitted through a cable of length l.[1] The impulse response of this cable for pulse transmission systems of high data rate is given by [21]:

$$h(t) = \frac{Al}{2\sqrt{\pi t^3}} e^{-\frac{(Al)^2}{4t}} \; ; \; t > 0 \tag{5.7}$$

with A given by

$$A = \frac{\sqrt{LC}}{2L} \tag{5.8}$$

which is related to the propagation constant γ. In (5.8) L is the inductance and C is the capacitance, both of the cable. Then, from (5.7) it is clear that the length of the cable affects $h(t)$. By considering five possible values of the product, $Al = 2, 3, 4, 5, 6$, the family of curves shown in Figure 5.6 emerges. As can be seen, the greater the factor Al, the larger the dispersion introduced in the pulse would be, which would limit the data rate. The increase of Al

---

[1]Although we are interested in the distortion introduced by an antenna of ultra-wide bandwidths, it is illustrative to begin the discussion with the transmission line, which, as the reader will remember, can be transformed into a radiator element, as noted in Chapter 2.

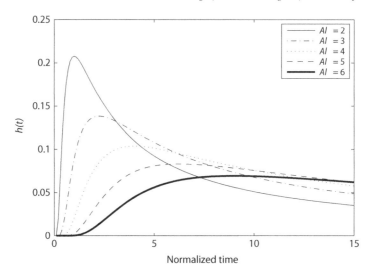

FIGURE 5.6
Impulse response of a cable for different lengths.

can be attributed either to a larger length l, or a greater value of A. Equation (5.8) states that A is a function of the inductance and the capacitance of the cable; as is well known, these parameters are affected by the geometry of the transmission line.

Now, a similar analysis could be applied to UWB antennas, provided that an antenna also physically presents certain capacitances and inductances, which depend on its structure (radiator shape and dimensions, feeder point, ground plane, etc.). Therefore, we can establish that the antenna impulse response is in general,

$$h(t) = f(t, \gamma) \qquad (5.9)$$

with γ the propagation constant depending on the antenna structure. The key point in terms of the UWB antenna is that the variables related to this structure are, in turn, a function of frequency, in such a way that diverse resonances could be generated at different electrical dimensions of the structure, so modifying the pulse shape.

It is worth noting here that the dispersion phenomenon can be considered as a particular type of distortion, where a pulse is stretched out into a longer waveform [6]. Although other forms of distortion could be presented (e.g., changes in the pulse shape, as shown in Section 5.4.1), dispersion is the first step in distortion analysis, provided it can introduce symbol interference and can limit the data rate.

5.4 Phase Linearity

5.4.1 Frequency response

In the frequency domain, it would be desirable for the antenna transfer function, $H(\omega)$, to present a flat or quasi-flat response on the frequency range of interest, in such a way that the signal spectrum at the antenna output resembles its corresponding input spectrum (see Figure 5.7 where the spectrum $Y(\omega)$ is equal to the spectrum $X(\omega)$ within the interval $\omega_L < \omega < \omega_H$). The meaning of a flat response of $H(\omega)$ in an interval $-\infty < \omega < \infty$ is that $h(t)$ is a delta function which, when convolved with the antenna input signal $x(t)$, produces a signal $y(t) = a\,x(t)$, where a is a constant.

Now, as the bandwidth is increased, the ideal situation described above is harder to achieve, such that a spectral modification is introduced; in fact, a UWB antenna is considered to behave as a shape filter. This behavior is explained in [22], where the transfer function and the gain response through the operation band are analyzed using three types of antennas for a UWB system (i.e., locating a pair of face-to-face antennas separated a certain distance, such that free space propagation conditions are presented). Strictly speaking, the antenna gain response does not correspond to $H(\omega)$, but through this response it can be determined. In this work, Lauber and Palaninathan define the pulse distortion as: "How faithfully does the received UWB pulse correspond to the transmitted pulse in shape?", and evaluate it based on the correlation coefficient obtained by comparing the received and the generated pulses [22]. Therefore, a non-flat gain response produces moderate or low correlation coefficients and consequently distortion of the signal, corresponding to the filtering effect.

As in the aforementioned reference, and in much other research, the pulse distortion has been addressed through the transfer function of the antenna.

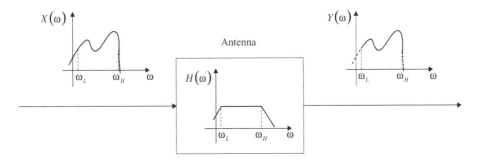

FIGURE 5.7
Flat transfer function of the antenna.

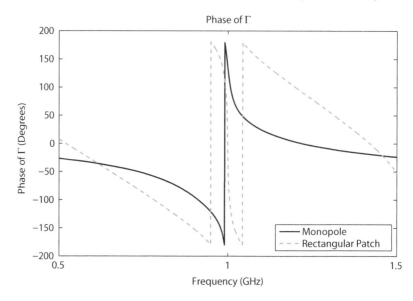

FIGURE 5.8
Simulated phase response for a wire monopole and a rectangular patch antenna.

Just as in the case of the reflection coefficient discussed in Chapter 2, this quantity can be represented by its magnitude $|H|$ and its phase $\phi(\omega)$:

$$H(\omega) = |H| \, e^{j\phi(\omega)} \qquad (5.10)$$

Note that in contrast to the expression of the reflection coefficient Γ for narrowband antennas given by Equation (2.5), now we are stressing the dependence of the phase as a function of frequency ω due to the wider bandwidth. Of course, this frequency dependence is applied to any phase response (including that corresponding to Γ). In order to illustrate this frequency dependance of the phase, Figure 5.8 depicts the phase response of Γ for the 1 GHz monopole wire and the rectangular patch antennas, whose reflection coefficient magnitude was shown in Figure 2.12. For comparison purposes, the phase response of a UWB planar monopole is shown in Figure 5.9, where a non-linear behavior can be observed.

5.4.2   Measures of the change of phase: Phase center and group delay

The phase center can be explained from the propagation mechanism of a wavefront. Let us consider the far field spherical wavefront depicted in Figure 5.10. The imaginary lines perpendicular to the wavefront make up a set of non-parallel lines, which appear to diverge from a common origin, which is what

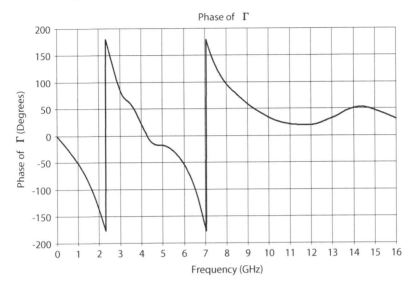

FIGURE 5.9
Simulated phase of the reflection coefficient for a UWB planar monopole.

in reflectors and lens antennas, is called the *phase center* [23]. This concept can of course be applied to other antenna structures. Thus, the phase center is referred to as "the effective origin of signals from an antenna" [6], and it "may be determined by a measurement of the phase of the feed radiation pattern as a function of angle or displacement" [24]. A fixed phase center will produce a linear phase response, and consequently the UWB pulses do not undergo distortion [19, 25].

The group delay is a measure of the time delay suffered by a UWB pulse, in proportion to the different wavelength dimensions of the antenna [26]. Since this quantity is a time measure, it is reasonable to anticipate that it relates to a change or velocity of another variable. Particularly, the factor that describes the above is the phase response, which is dependent on the frequency. Thus, the group delay is mathematically expressed by

$$\tau_g(\omega) = -\frac{d\phi(\omega)}{d\omega} \tag{5.11}$$

and its mean value over the band of interest is delimited by ω_L and ω_H

$$\overline{\tau}_g(\omega) = \frac{1}{\omega_L - \omega_H} \int_{\omega_L}^{\omega_H} \tau_g(\omega) d\omega \tag{5.12}$$

By analyzing (5.11), one can deduce that if $\phi(\omega)$ is linear, there will be a constant group delay. This is a necessary condition in order to have a

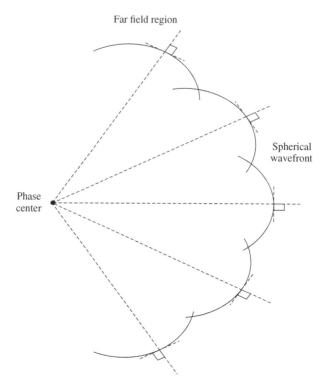

FIGURE 5.10
Representation of the phase center concept.

non-distorted pulse [7], otherwise a non-linear behavior of $\phi(\omega)$ will make that device present a potentially resonant character, implying that the structure can store energy, thus increasing the Q factor and as a consequence, reducing its bandwidth. This situation generates ringing and oscillations on the impulse response $h(t)$ of the antenna [7].

5.4.3   Phase response and pulse distortion

In many texts, study of phase linearity is based on the transfer function of a UWB system, where $H(\omega)$ corresponds to the channel response, instead of the antenna response as was previously discussed (see Section 5.4.1). Nevertheless, the phase behavior can also be studied based on the reflection coefficient of the antenna, because this quantity provides the phase response as well. Examples of this way of evaluating the phase response can be found in [12, 14, 20, 26, 27]. Now, according to the group delay concept explained above (Section 5.4.2), if this were constant, the source pulse would simply pass through the antenna, producing a delayed pulse but conserving its shape. A non-constant group delay response, then, will introduce not only a certain time delay, but also shape modifications in the pulse. Therefore, if the phase $\phi(\omega)$ presents a non-linear response, $\tau_g(\omega)$ will be not constant, so pulse distortion will be generated. From the above discussion, it is clear that UWB antennas must present a linear phase response in order to avoid pulse distortion. This relationship between linear phase and pulse distortion in UWB antennas has been stressed by several authors [7, 28, 29]. For instance, Kwon [29] points out that it is highly desirable that UWB antennas have a constant response both for the gain and the group delay through the operational frequency band, such that the received impulse has an appearance as close as possible to the impulse generated at the source.

5.5   Non-Linear and Quasi-Linear Phase Antennas

5.5.1   Frequency independent antennas

As mentioned in Chapter 1, although the increasing demand for wider and wider bandwidths is not a new subject, it has became imperative in recent years, given its current applications in diverse areas of modern life. Early attempts at achieving wide bandwidths in the 1960s were based on the so-called *frequency independent antennas*, whose operation principle is based on the fact that different antenna elements radiate at distinct frequencies. We can cite, for example, the well known log-periodic antenna, widely used for TV broadcasting reception. Figure 5.11 shows the performance of this type of antenna in terms of the reflection coefficient magnitude, and where we can see several resonances that allow different TV channels to be tuned by the reception apparatus.

However, it is exactly this operation principle that places a constraint on classical frequency independent antennas if they are to be used in pulsed applications like UWB systems. The fact that different elements of the antenna structure serve to radiate at distinct frequencies means that its phase center is not fixed, which in turn means that the phase of its reflection coefficient has a non-linear behavior (see for example the reflection coefficient phase of a 7-elements log-periodic antenna at 680 MHz depicted in Figure 5.12). As was discussed in Section 5.4.3, this antenna response has as a consequence for the pulse distortion (in essence, radiating elements produce a larger pulse duration, due to the increase in envelope width, τ_{FWHM} [7]).

Another example of an antenna that introduces distortion is the spiral antenna. Schantz shows how a pulse is dispersed (around twice its original duration) when it is transmitted using this antenna [30].

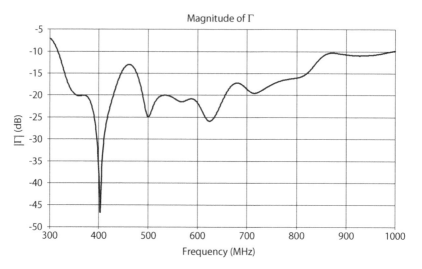

FIGURE 5.11

Magnitude of the reflection coefficient of a 7-elements 680 MHz log-periodic antenna.

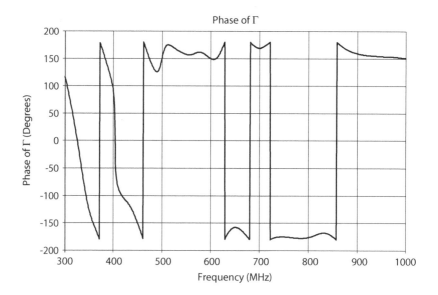

FIGURE 5.12

Phase response of the reflection coefficient of a 7-elements 680 MHz log-periodic antenna.

TABLE 5.1

Envelope width for different antennas [7]

Antenna	τ_{FWHM} (ps)
Vivaldi	135
Bowtie	140
Archimedian spiral	290
Log-periodic	805
Monocone	75

5.5.2 Other antennas

In principle, some antenna designs could be modified in order to avoid the pulse distortion effect. Ghosh et al., for instance, propose the inclusion of a resistive load profile in dipole, bicone, TEM horn, spiral and log-periodic antennas, in such a way that profile electromagnetically compensates for the effects of the discontinuities of their structures, like reflections that could form a standing wave [13]. As Ghosh et al. point out, the idea is to focus the energy of a single pulse to obtain a traveling wave. However, this enhancement in terms of pulse distortion reduction produces degradation of antenna efficiency (between 50−60%, whereas non-loaded antennas achieve an efficiency of 80−90% or more [13]).

The work of Wiesbeck et al. is an effort to address the time domain properties of some UWB antennas and other potential candidates for ultra wideband, by trying to maintain a constant input impedance [7]. In this article, the envelope width of the transient response (see Section 5.2 above) is used as the reference parameter to get a comparison. Specifically, the antennas under test are Vivaldi, bowtie, archimedian spiral, log-periodic, and monocone, and whose results are given in Table 5.1.

As can be appreciated, the log-periodic antenna presents the largest envelope width, with a value exceeding the few hundred picoseconds suggested [7], confirming its non-linear phase feature mentioned above. The bowtie antenna was also studied by Yazdandoost et al., who report that this device has a nearly linear phase response, implying a constant group delay and therefore no distortion in the pulse shape [12]. The Vivaldi antenna is an interesting case, studied by several authors. This antenna is usually considered to be a non-dispersive antenna, due to its practically linear phase feature. Although, according to values shown in Table 5.1, it has an envelope with of $h(t)$ – comparable to that of the bowtie antenna (associated to multiple reflections and parasitic currents at the substrate edges [7]) – the Vivaldi antenna represents one of the most promising structures for UWB applications.

Other linear phase antennas reported in the literature include the semi-elliptic slot antenna [20], the planar Penta-Gasket-Koch fractal antenna (with small phase distortions in the lowest segment of the 2-20 GHz bandwidth) [14], and the rectangular patch with a slot bowtie on it [27]. It is worth noting that the objective here is not to discuss the wide variety of research findings relating to the phase response of different antennas. Instead, through Chapters 6 and 7, some simulations and experimental results are analyzed for omnidirectional and directive UWB antennas, where both the magnitude and the phase of Γ are presented. Based on these results, comments are made concerning their dispersive or non-dispersive features.

Bibliography

[1] S. S. Haykin and B. Van Veen. *Signals and Systems.* John Wiley & Sons, New York, 1999.

[2] J. G. Proakis and D. G. Manolakis. *Digital Signal Processing: Principles, Algorithms and Applications.* Macmillan, New York, 2nd edition, 1992.

[3] B. P. Lathi. *Signals, Systems and Communication.* John Wiley & Sons, New York, 1965.

[4] K. S. Shanmugan. *Digital and Analog Communication Systems.* John Wiley & Sons, New York, 1979.

[5] FCC. First report and order, revision of part 15 of the commission's rules regarding ultra-wideband transmission systems. Technical report, Federal Communications Commission, 2002.

[6] H. Schantz. *The Art and Science of Ultra Wideband Antennas.* Artech House, Norwood, MA, 2005.

[7] W. Wiesbeck, G. Adamiuk, and C. Sturm. Basic properties and design principles of UWB antennas. *Proceedings of the IEEE*, 97(2):372–385, 2009.

[8] Z. N. Chen, X. H. Wu, H. F. Li, N. Yang, and M. Y. W. Chia. Considerations for source pulses and antennas in UWB radio systems. *IEEE Transactions on Antennas and Propagation*, 52(7):1739–1748, 2004.

[9] M. Welborn and J. McCorkle. The importance of fractional bandwidth in ultra-wideband pulse design. In *IEEE International Conference on Communications*, number 2, pages 753–757, 2002.

[10] G. Lu, P. Spasojevic, and L. Greenstein. Antenna and pulse designs for meeting UWB spectrum density requirements. In *2003 IEEE Conference on Ultra Wideband Systems and Technologies*, pages 162–166, 2003.

[11] D. M. Pozar. Waveform optimizations for ultrawideband radio systems. *IEEE Transactions on Antennas and Propagation*, 51(9):2335–2345, 2003.

[12] K. Y. Yazdandoost, H. Zhan, and R. Kohno. Ultra-wideband antenna and pulse waveform for UWB applications. In *6th International Conference on ITS Telecommunications*, pages 345–348, 2006.

[13] D. Ghosh, A. De, M. C. Taylor, T. K. Sarkar, M. C. Wicks, and E. L. Mokole. Transmission and reception by ultra-wideband (UWB) antennas. *IEEE Antennas and Propagation Magazine*, 48(5):67–99, 2006.

[14] M. Naghshvarian-Jahromi. Novel wideband planar fractal monopole antenna. *IEEE Transactions on Antennas and Propagation*, 56(2):3844–3849, 2008.

[15] G. F. Ross. A time domain criterion for the design of wideband radiating elements. *IEEE Transactions on Antennas and Propagation*, 16(3):355–356, 1968.

[16] A. Shlivinski, E. Heyman, and R. Kastner. Antenna characterization in the time domain. *IEEE Transactions on Antennas and Propagation*, 45(7):1140–1149, 1997.

[17] T. W. Hertel and G. S. Smith. On the dispersive properties of the conical spiral antenna and its use for pulsed radiation. *IEEE Transactions on Antennas and Propagation*, 51(7):1426–1433, 2003.

[18] W. Kong, Y. Zhu, and G. Wang. Effects of pulse distortion in UWB radiation on UWB impulse communications. In *International Conference on Wireless Communications, Networking and Mobile Computing*, volume 1, pages 344–347, 2005.

[19] J. S. McLean and R. Sutton. UWB antenna characterization. In *Proceedings of the 2008 IEEE International Conference on Ultra-wideband*, volume 2, pages 113–116, 2008.

[20] M. Gopikrishna, D. Das Krishna, C. K. Anandan, P. Mohanan, and K. Vasudevan. Design of a compact semi-elliptic monopole slot antenna for UWB systems. *IEEE Transactions on Antennas and Propagation*, 57(6):1834–1837, 2009.

[21] *Transmission Systems for Communications*. Bell Telephone Laboratories, Wiston-Salem, North Carolina, 4th edition, 1971.

[22] W. Lauber and S. Palaninathan. Ultra-wideband antenna characteristics and pulse distortion measurements. In *The 2006 IEEE International Conference on Ultra-Wideband*, pages 617–622, 2006.

[23] L. V. Blake. *Antennas*. Artech House, 1984.

[24] W. L. Weeks. *Antenna Engineering*. McGraw-Hill, New York, 1968.

[25] K. Y. Yazdandoost and R. Kohno. Ultra wideband antenna. *IEEE Communications Magazine*, 42(6):S29–S32, 2004.

[26] P. McEvoy, M. John, S. Curto, and M. J. Ammann. Group delay performance of ultra wideband monopole antennas for communication applications. In *2008 Loughborough Antennas and Propagation Conference*, pages 377–380, 2008.

[27] H. Zhang, X. Zhou, K. Y. Yazdandoost, and I. Chlamtac. Multiple signal waveforms adaptation in cognitive ultra-wideband radio evolution. *IEEE Jorunal on Selected Areas in Communications*, 24(4):878–884, 2006.

[28] D. D. Wentzloff, R. Blázquez, F. S. Lee, B. P. Ginsburg, J. Powell, and A. P. Chandrakasan. System design considerations for ultra-wideband considerations. *IEEE Communications Magazine*, 43(8):114–121, 2005.

[29] D. H. Kwon. Effect of antenna gain and group delay variations on pulse-preserving capabilities of ultrawideband antennas. *IEEE Transactions on Antennas and Propagation*, 54(8):2208–2215, 2006.

[30] H. G. Schantz. Dispersion and UWB antennas. In *2004 International Worshop on Ultrawideband Systems and Technologies*, pages 161–165, 2004.

6

Design of Omnidirectional UWB Antennas for Communications

CONTENTS

6.1   Introduction to Omnidirectional UWB Antennas

As was shown in Chapter 3, many UWB antennas have been designed to have ominidirectional radiation patterns. However, like other antenna parameters, conserving this feature with low variations over the operational ultra wideband bandwidth is a difficult task. In Chapter 2 we explained that the directionality (which is related to the gain) of an antenna is determined by its radiation pattern, provided it describes the spatial distribution of energy radiated or received. Thus, it is desirable that the gain is constant or quasi-constant for a wide range of frequencies.

The classical omnidirectional narrowband antenna is the infinitesimally small dipole, an ideal model of which is used at the beginning of many antenna theory books. Unfortunately, the narrowband behavior of this antenna is not suitable for using it in the UWB field, and it is recommended only as a preliminary theoretical approach. The ideal dipole antenna used in narrowband antenna theory is replaced in UWB by a cylindrical monopole antenna, because its radiation behavior is similar to the infinitesimally small dipole, but with a wider frequency band (remember the solid-planar principle discussed in Chapter 4). Moreover, the cylindrical monopole antenna was used to find an equation to determine the lower cut-off frequency of the operational bandwidth of omnidirectional UWB planar antennas.

Many of the omnidirectional UWB antennas presented in Chapter 3 were designed using high frequency structure simulators, whose design methodology has been omitted. A design methodology is explained in this chapter to develop omnidirectional UWB antennas for diverse operational bandwidths with different grades of variation in their radiation pattern, depending on the specific requirements. This design flexibility is achieved first of all because it is possible to determine the height of the radiator from an equation that requires only the lower cut-off frequency. Also, by following the idea that one of the dimensions of the antennas must be longer (see Section 4.3), volumetric antennas are explored to achieve better radiation pattern features with respect to planar antennas. Therefore, a pseudo-volumetric omnidirectional UWB antenna is considered as an example to conserve the omnidirectionality of the radiation pattern in UWB planar antennas.

Although in this book the operational bandwidth of a UWB antenna is first evaluated in terms of the matching impedance, and second for the character of its phase, variations of the radiation pattern as a function of frequency are also important. Depending on particular design requirements, it can be used to define the operational bandwidth of an omnidirectional UWB antenna (in fact, one of the definitions of bandwidth for any type of antenna is related to radiation pattern behavior, as was explained in Chapter 2). The possibility of choosing a different physical parameter to define antenna bandwidth follows the IEEE 145-1993 standard [1].

Due to its versatility, a planar monopole antenna (PMA) has been chosen as a starting point to study the omnidirectional characteristic of UWB antennas. PMA can adopt multiple shapes including circular, triangular, square, etc. A square shape is chosen here, because its radiation pattern has less degradation compared to other shapes [2].

On the other hand, and for comparison purposes, the initial dimensions for the PMA design, based on a methodology discussed below in Section 6.3, are similar to those reported in [3]. At this point, it is therefore necessary to analyze the behavior of a square planar monopole antenna (SPMA) without any impedance matching technique.

Once this initial analysis has been done, the design methodology for tuning a rectangular UWB planar monopole antenna (RPMA) will be established.

Moreover, to improve the omnidirectional radiation pattern and its stability at higher frequencies, the single radiator design can be upgraded to an orthogonal or $\pi/4$ bi-orthogonal antenna design.

As regards planarized structures, it is possible to use a similar design methodology. As is known, this antenna has the radiator and the ground plane printed on the same board, which reduces its physical dimensions, making it suitable for mobile devices.

The antennas' performance studied here is evaluated in terms of impedance matching, phase response and radiation pattern. It is worth noting that the phase behavior is analyzed through the corresponding response of the reflection coefficient. As was mentioned in Chapter 1, all simulations presented in this chapter have been carried out using the software CST Microwave Studio. The measured results of Γ are carried out with an Agilent network analyzer model E8362B.

6.2   Starting Point: Monopole Square Planar Antenna

In order to find a starting point for designing omnidirectional planar antennas, let us consider a square planar antenna, whose radiation pattern is typically omnidirectional. Then, we need to test if this antenna fulfills two fundamental requirements for UWB. First, it must be matched to the bandwidth specified for UWB applications, as explained in Chapter 4. Second, the antenna must have a linear or quasi-linear phase characteristic, in order to avoid the pulse distortion effect described in Chapter 5.[1] Thus, based on the work of Su et al. [3], we can consider the design shown in Figure 6.1.

The simulated radiation pattern for this antenna is presented in Figure 6.2 for 3, 11, 12 and 14 GHz. As can be seen, the PMA with the dimensions illustrated in Figure 6.1 presents an omnidirectional radiation pattern through the simulated frequency band. However, in terms of the matching and phase characteristics, the antenna depicted in Figure 6.1 does not fulfill either criterion of UWB as shown in Figures 6.3 and 6.4, of simulated magnitude and phase of the reflection coefficient, respectively. Remember that the matching threshold is set to be $|\Gamma| < -10\,\mathrm{dB}$.

Due to the fact that the current dimensions of this antenna do not provide the necessary conditions to be matched in the UWB band, an impedance matching technique must be adopted like those mentioned in Chapter 4. Therefore, it is necessary to follow a certain design methodology, which is discussed in the following section.

---

[1]It is worth noting that the requirement of a linear response of the phase can be relaxed if the antenna is not to be used for high-rate communications applications. Spectrum monitoring can be a possible niche for non-linear phase UWB antennas anyway.

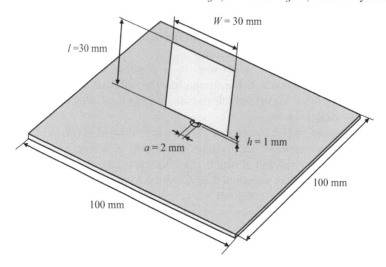

FIGURE 6.1
Geometry and dimensions of a square monopole planar antenna.

6.3   Design Methodology for Planar Structures

The design methodology to be explained below is similar to that reported in [4], but with a more detailed process. Originally it was proposed in [5] and applied in [6], in seeking to match the antenna to the UWB band, and to try and reduce variations in the radiation pattern as a function of the frequency. Essentially, this methodology is based on the knowledge of the lower cut-off frequency f_L determined by Equation (4.17) (see Section 4.7), and considers four variables, which are different dimensions of the antenna: the bevel angle ψ, radiator width W, the feed width a and the height between the ground plane and the radiator h (see Figure 6.5, where a rectangular planar antenna with a bevel on both sides of its bottom is included).

The central idea of the design methodology is that each of the aforementioned variables is varied individually until an "optimal" value is found. It is important to point out here that although each variable experiences a variation of N times (let's say N_x, where the subindex x stands for the involved variable ψ, W, a or h), the individual "optimal" values are not found in a simultaneous process. Indeed, this methodology follows a serial flow as described below (see the flow diagram depicted in Figure 6.6):

1. Initiate with a certain values of ψ_0, W_0, a_0 and h_0.

2. Evaluate $\Gamma(\psi_i)$ for $i = 1, \ldots, N_\psi$ (N_ψ a positive integer) and select an "optimal" curve for that $\Gamma(\psi_{opt})$, interchange ψ_0 by ψ_{opt}.

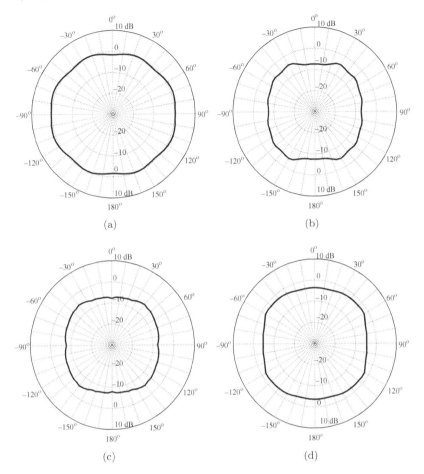

FIGURE 6.2
Simulated radiation pattern of the square monopole planar antenna at different frequencies: (a) 3 GHz, (b) 11 GHz, (c) 12 GHz, (d) 14 GHz.

3. Check if the "optimal" curve for $\Gamma(\psi_{\mathrm{opt}})$ fulfills a UWB criterion.
   If so, the method concludes.

4. If the UWB criterion is not fulfilled, evaluate $\Gamma(W_j)$ for $j = 1, \ldots, N_W$ (N_W a positive integer) and select an "optimal" curve
   for that $\Gamma(W_{\mathrm{opt}})$, interchange W_0 by W_{opt}.

5. Check if the "optimal" curve for $\Gamma(W_{\mathrm{opt}})$ fulfills a UWB criterion.
   If so, the method concludes.

6. If the UWB criterion is not fulfilled, evaluate $\Gamma(a_k)$ for $k = 1, \ldots, N_a$ (N_a a positive integer) and select an "optimal" curve
   for that $\Gamma(a_{\mathrm{opt}})$, interchange a_0 by a_{opt}.

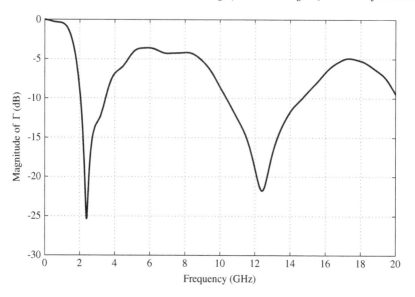

FIGURE 6.3

Simulated reflection coefficient magnitude of the square monopole planar antenna.

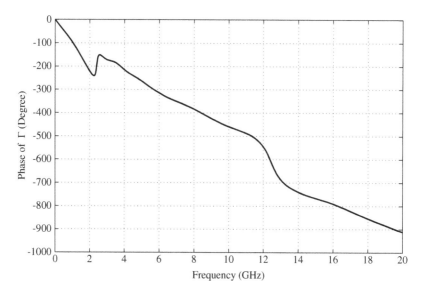

FIGURE 6.4

Simulated reflection coefficient phase of the square monopole planar antenna.

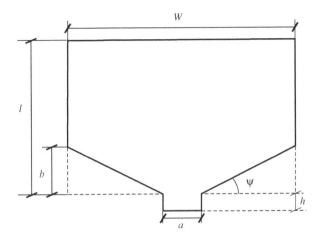

FIGURE 6.5
Variables involved in the design methodology of a UWB antenna.

7. Check if the "optimal" curve for $\Gamma(a_{\text{opt}})$ fulfills a UWB criterion. If so, the method concludes.

8. If the UWB criterion is not fulfilled, evaluate $\Gamma(h_l)$ for $l = 1, \ldots, N_l$ (N_l a positive integer) and select an "optimal" curve for that $\Gamma(h_{\text{opt}})$, interchange h_0 by h_{opt}.

9. Check if the "optimal" curve for $\Gamma(h_{\text{opt}})$ fulfills a UWB criterion. If so, the method concludes.

10. If the UWB criterion is not fulfilled, begin step (2) as long as the maximum number of iterations had been not reached.

The evaluation of Γ using each variable ψ, W, a and h can vary depending on the designer (for this reason, the maximum number of times that each evaluation is carried out is identified with distinct letters N_ψ, N_W, N_a and N_h, respectively). The idea of using the variables ψ, W, a and h as parameters to evaluate antenna performance is based on the fact that they modify current distribution on the radiator, as well as its capacitances and inductances [5], in such a way that the stored and radiated energy is altered, and consequently the bandwidth is also modified. Some simulation examples that show antenna performance by varying these parameters are illustrated in Section 6.4.

It is worth noting that the evaluation mechanism of $\Gamma(x)$ consists of calculating some of its statistical values. In particular, we use the median and the percentile range of $|\Gamma(x)|$, which means selecting the value of the parameter under evaluation that provides the best trade-off between the median value that provides the best matching coefficient, and the range of $|\Gamma(x)| < -10\,\text{dB}$ with the largest percentage. For each of the variables under evaluation, the

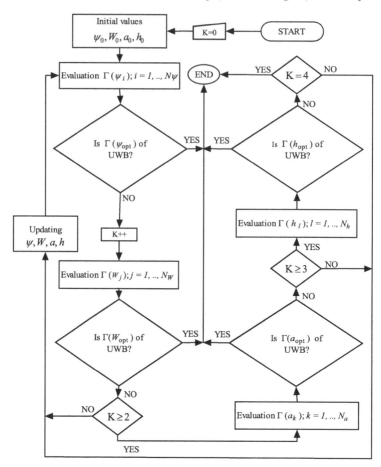

FIGURE 6.6
Flow diagram of the design methodology for an omnidirectional planar antenna.

values resulting from this process are labeled as ψ_{opt}, W_{opt}, a_{opt} and h_{opt}. Figure 6.7 illustrates the flow diagram of the general evaluation process.

6.4   Simulation Results Based on the Design Methodology for Planar Structures

The design methodology explained in Section 6.3 is used below in order to describe the effects that each variable concerned has on antenna performance.

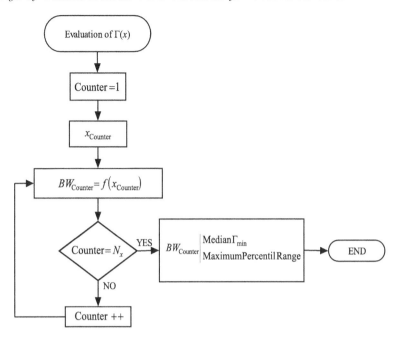

FIGURE 6.7
General flow diagram of the evaluation of $\Gamma(x)$.

The order as each variable is evaluated follows the aforementioned design methodology. Thus, the original dimensions of the antenna are the same as in Figure 6.1.

6.4.1 Variation of beveling angle

The bevel on the antenna used as a matching technique [7] was explained in Section 4.8. We will consider three possible bevel angles. Also consider that the radiator has a base of 14 mm and for $a = 2$ mm and $W = 30$ mm, and from a simple relation between the angle ψ and the dimension b, three angles are considered for illustration purposes $\psi = 0, 32.74°, 35.53°$. Figure 6.8 shows the simulated results for the coefficient reflection magnitude. As can be seen, when $\psi \neq 0$, the pair of curves corresponding to $\psi = 32.74°$ and $\psi = 35.53°$ move downwards, so tending to be in a matched region. Although both plots are very close each other, a certain mismatching can be noted for $\psi = 35.53°$. This means that the angle $\psi = 32.74°$ provides better matching conditions, and hence this variable must not be increased indefinitely.

Although the current configuration of the SPMA does not provide matching conditions at all, let us present the effect associated with the bevel angle that is introduced in antenna performance in terms of its radiation pattern.

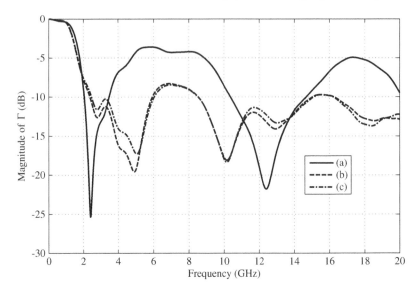

FIGURE 6.8

Simulated reflection coefficient magnitude of the SPMA for three bevel angles: (a) $\psi = 0°$, (b) $\psi = 32.74°$ and (c) $\psi = 35.54°$.

Figure 6.9 illustrates the radiation pattern for different frequencies for the SPMA with $\psi = 32.74°$. As can be seen, the radiation pattern almost conserves its omnidirectionality.

6.4.2   Variation of the radiator width

The next step of the design methodology consists of varying the width of the antenna radiator, W. Thus, there is a height/width ratio different to the unity, and the radiator now has a rectangular shape. Let us assume a set of possible radiator widths, from $W = 32$ mm to $W = 60$ mm in steps of 2 mm. By simulating the reflection coefficient magnitude with $\psi = 32.74°$, a set of curves is obtained, from which that for $W = 50$ mm is selected due to its performance. Figure 6.10 shows this plot; the plots of $|\Gamma|$ for the original $W = 30$ mm square antenna with $\psi = 0°$ and $\psi = 32.74°$ are also included for comparison purposes. As can be seen, the plot of the rectangular planar antenna is moved downwards from the threshold of -10 dB from 2.46 GHz (although there is a critical point around 8 GHz), providing a better performance in terms of impedance matching.

Regarding the radiation pattern, let us examine it through the results shown in Figure 6.11. As can be seen, this antenna configuration produces a certain directionality in the radiation pattern, particularly as the frequency is increased.

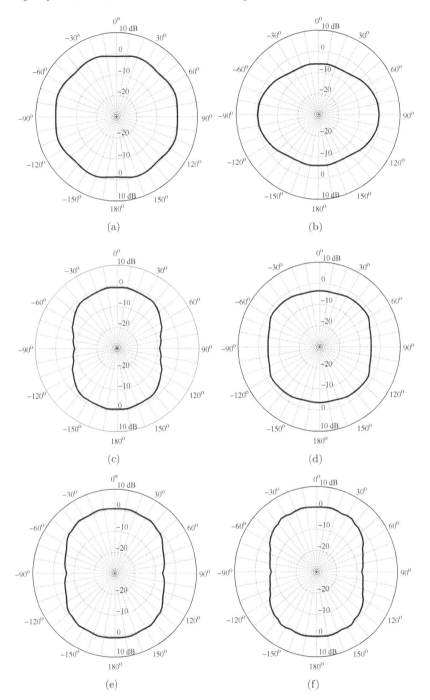

FIGURE 6.9
Simulated radiation pattern for the SPMA with $\psi = 32.74°$ at (a) 3 GHz, (b) 6 GHz, (c) 9 GHz, (d) 12 GHz, (e) 15 GHz and (f) 18 GHz.

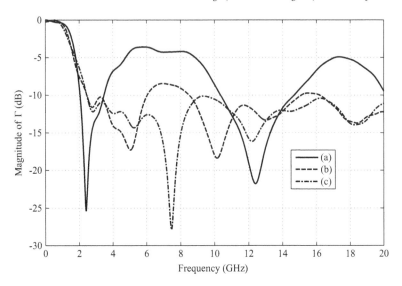

FIGURE 6.10
Simulated reflection coefficient magnitude for the $W = 30$ mm square planar antenna with (a) $\psi = 0°$ and (b) $\psi = 32.74°$, and for the rectangular planar antenna with a height/width ratio of 0.6 ($W = 50$ mm) for $\psi = 20.6°$ (c).

It is worth noting here that variations of the radiator width modify the bevel angle (see Figure 6.5) and therefore, one can now simulate the 50 mm width rectangular antenna for different bevel angles, according to the design methodology. For example, Figure 6.12 shows the simulation results of $|\Gamma|$ corresponding to four bevel angles. From these results, the angle $\psi = 20.6°$ can be chosen due to the fact that it presents the best statistical characteristics for UWB (median of $|\Gamma| = -12.35$ dB and a percentile range of 88.4% for $|\Gamma| < -10$ dB).

6.4.3   Variation of the feeder width

Let us now evaluate the impact of the feeder width, a, on the antenna performance. In order to follow the design methodology of Section 6.3, consider the rectangular antenna dimensions achieved up to this point ($\psi = 20.6°$ and $W = 50$ mm). Figure 6.13 depicts the simulated reflection coefficient magnitude of this antenna for six widths of the feed. As can be seen, the wider the feed width, the better the achieved matching. This is translated into a shift of f_L (from 2.44 GHz for $a = 2$ mm to 2.31 GHz for $a = 4$ mm) and into a displacement of the curve downward, which contributes to an increase in the bandwidth. However, this tendency is not conserved in all cases, because when $a = 4$ mm the curve starts to move upward near the threshold of -10 dB causing a mismatch around 17 GHz. Hence $a = 3.6$ mm can be considered as an acceptable value.

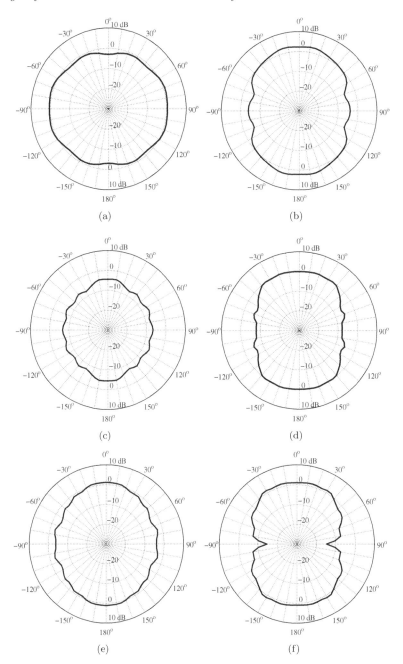

FIGURE 6.11
Simulated radiation pattern for the rectangular planar antenna with a height/width ratio of 0.6 ($W = 50$ mm) at (a) 3 GHz, (b) 6 GHz, (c) 9 GHz, (d) 12 GHz, (e) 15 GHz and (f) 18 GHz.

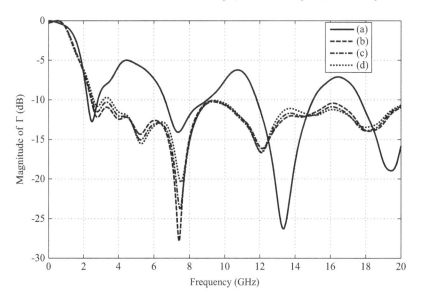

FIGURE 6.12
Simulated reflection coefficient magnitude of the RPMA of $W = 50\,\text{mm}$ and for four bevel angles: (a) $\psi = 2.4°$, (b) $\psi = 20.6°$, (c) $\psi = 22.6°$ and (d) $\psi = 24.6°$.

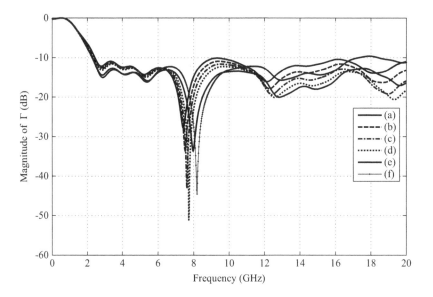

FIGURE 6.13
Simulated reflection coefficient magnitude of the RPMA of $W = 50\,\text{mm}$ and $\psi = 20.6°$ for six widths of the feeder: (a) $a = 2.0$ mm, (b) $a = 2.4$ mm, (c) $a = 2.8$ mm, (d) $a = 3.2$ mm, (e) $a = 3.6$ mm and (f) $a = 4.0$ mm.

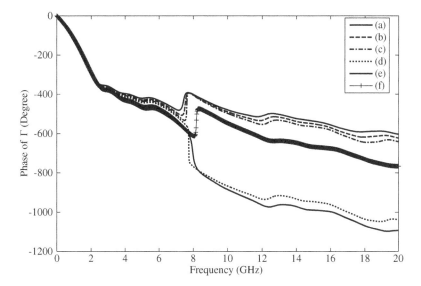

FIGURE 6.14

Simulated reflection coefficient phase of the RPMA of $W = 50$ mm and $\psi = 20.6°$ for six widths of the feeder: (a) $a = 2.0$ mm, (b) $a = 2.4$ mm, (c) $a = 2.8$ mm, (d) $a = 4.0$ mm, (e) $a = 3.6$ mm and (f) $a = 3.2$ mm.

Regarding the phase characteristic, Figure 6.14 depicts the results obtained via CST Microwave Studio. As can be seen, the phase still has a non-linear characteristic for all simulated cases. It is worth noting that the phase for $a = 3.6$ mm presents less abrupt changes, which is beneficial in reducing possible pulse distortion.

As the design methodology indicates, the different variables can be adjusted as many times as the designer considers that the antenna provides an acceptable UWB performance. Thus, after different variations of ψ, W and a, the original PMA is tuned with the dimensions depicted in Figure 6.15 and whose magnitude and phase of Γ obtained by simulations are shown in Figures 6.16 and 6.17, respectively. As can be seen from these figures, the dimensions and geometry of the rectangular planar antenna offer an excellent matching response, and an almost smooth (although not linear) phase transition.

In terms of the radiation pattern, Figure 6.18 shows the variation that it experiences through the frequency range over which simulations were conducted. According to results shown in this figure, although no notable differences are noted compared to previous configurations, the loss of omnidirectionality is still present in this design as the frequency increases.

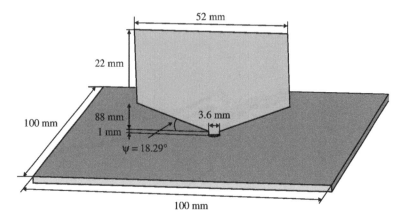

FIGURE 6.15
Geometry of the rectangular planar monopole antenna with a height/width ratio of 0.58, $\psi = 18.29°$ and $a = 3.6$ mm.

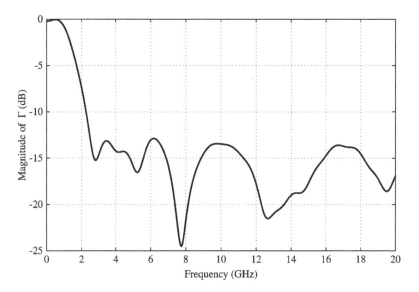

FIGURE 6.16
Simulated magnitude of the reflection coefficient for the RPMA with $\psi = 18.29°$, $W = 52$ mm and $a = 3.6$ mm.

6.4.4 Variation of the height between the ground plane and the radiator

First consider the antenna geometry of Figure 6.15, over which the height between the ground plane and the radiator, h, is the parameter to be varied as

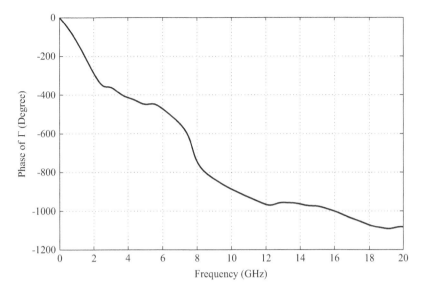

FIGURE 6.17
Simulated reflection coefficient phase for the RPMA with $\psi = 18.29°$, $W = 52\,\text{mm}$ and $a = 3.6\,\text{mm}$.

the design methodology indicates. Then, six values are arbitrarily considered for h, 1.0, 1.1, 1.2, 1.3, 1.4 and 1.5 mm. The obtained results are shown in Figures 6.19 and 6.20 for the magnitude and phase of the simulated reflection coefficient, respectively.

As can be seen, as the value of h is increased, the impedance matching is reduced, so affecting the bandwidth. The best trade-off for this variable is for $h = 1\,\text{mm}$ with a median of $|\Gamma| = -14.70\,\text{dB}$ and a percentile range for $|\Gamma| < -10\,\text{dB}$ of 89.1%. In terms of the phase response, it conserves its non-linear characteristic, which is similar for almost all values of h, except for $h = 1.5\,\text{mm}$, where the phase curve presents a severe transition around 8 GHz, which is related to a resonance frequency.

6.5 Reduction of Variations in the Radiation Pattern

As has been seen above, conserving the radiation pattern shape for a large bandwidth is a complicated task. Therefore, the use of more than one orthogonal radiating element has been considered as a mechanism by which to reduce the variations of the radiation pattern of a planar UWB antenna [8] and [6]. The principle of this idea is that the electromagnetic fields generated in each

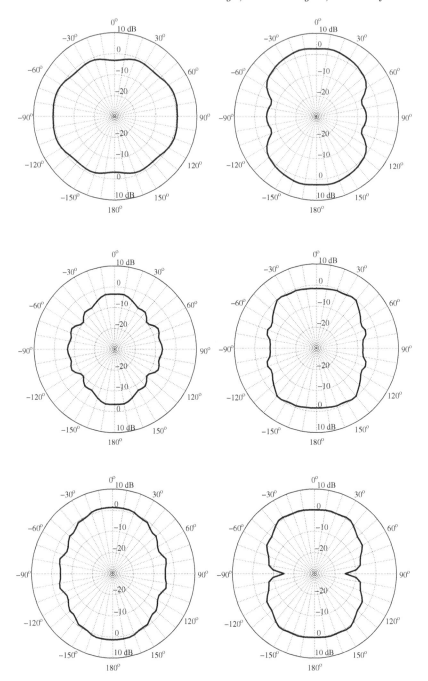

FIGURE 6.18

Radiation pattern for the RPMA with $\psi = 18.29°$, $W = 52\,\text{mm}$ and $a = 3.6\,\text{mm}$ at (a) 3 GHz, (b) 6 GHz, (c) 9 GHz, (d) 12 GHz, (e) 15 GHz and (f) 18 GHz.

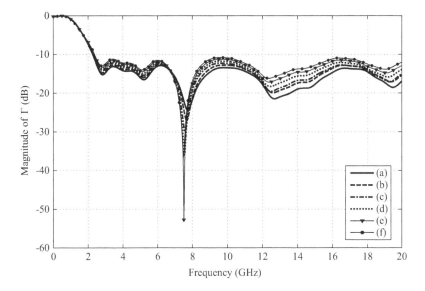

FIGURE 6.19
Simulated magnitude of the reflection coefficient for the RPMA with $\psi = 18.29°$, $W = 52$ mm, $a = 3.6$ mm and for different values of h: (a) $h = 1.0$ mm, (b) $h = 1.1$ mm, (c) $h = 1.2$ mm, (d) $h = 1.3$ mm, (e) $h = 1.4$ mm and (f) $h = 1.5$ mm.

element of the composed radiator are added, which resembles the behavior of a volumetric structure. Then, the electromagnetic field generated with two or more orthogonal elements reduces the frequency-dependent character of the radiation pattern in planar monopole antennas.

Let us assume the bi-orthogonal structure of four planar elements shown in Figure 6.21, where each rectangular element is 45° shifted and has been tuned following the design methodology explained in Section 6.3 [6]. The corresponding simulated reflection coefficient magnitude and radiation pattern are depicted in Figures 6.22 and 6.23, respectively. As can be seen, this radiator is well matched, although a critical frequency is present at 10.78 GHz, where $|\Gamma| = -9.98$ dB. In terms of radiation pattern, better omnidirectional characteristics are observed if we compare results of Figure 6.23 and those of single element radiator (see for example Figure 6.18).

The reader may think, moreover, that in the limit case where the number of elements is infinity, the best performance of the antenna would be attained. This notion was analyzed by Peyrot-Solis et al. in [6], where it was demonstrated that although the radiation pattern shape could be conserved, the larger the number of elements, the narrower the bandwidth in terms of the impedance matching. This fact can be analyzed through the coefficient reflection magnitude of a revolution-solid antenna and an orthogonal radiator (see [6] for design details of both antennas) as is illustrated in Figure 6.22. As

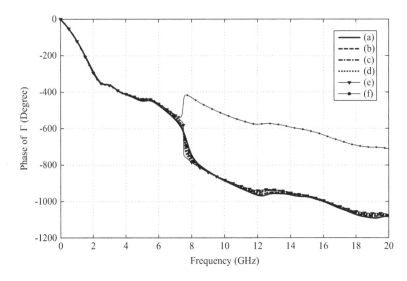

FIGURE 6.20
Simulated phase of the reflection coefficient for the RPMA with $\psi = 18.29°$, $W = 52$ mm, $a = 3.6$ mm and for different values of h: (a) $h = 1.0$ mm, (b) $h = 1.1$ mm, (c) $h = 1.2$ mm, (d) $h = 1.3$ mm, (e) $h = 1.4$ mm and (f) $h = 1.5$ mm.

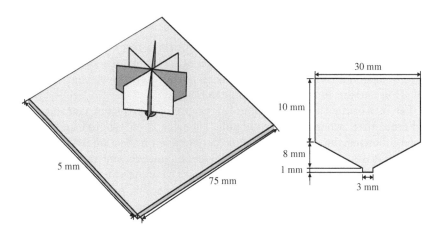

FIGURE 6.21
Geometry and dimensions of a bi-orthogonal UWB antenna.

can be seen, the upper cut-off frequency is reduced from almost 16 GHz for the orthogonal antenna to 8 GHz for the revolution-solid antenna. This bandwidth reduction is attributed to the increment of the capacitances associated with the infinite number of elements, which affect the resonance frequency at

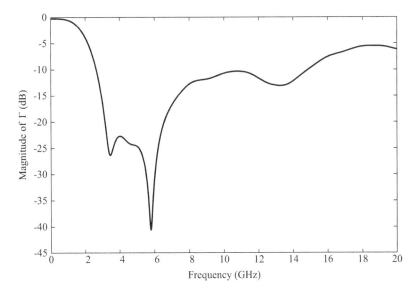

FIGURE 6.22
Reflection coefficient magnitude for the antenna of Figure 6.21.

around 14 GHz. Observe that for the revolution-solid antenna its $|\Gamma|$ curve tends to move upward, mainly at high frequencies, which is transformed to an impedance mismatching.

6.6   Design of Planarized UWB Antennas

6.6.1   General concepts

Let us now introduce some guidelines for the design of planarized UWB antennas with omnidirectional radiation properties, which provide a wider bandwidth than that reported in the literature (e.g., [9]). The design methodology in this case is a simplified version of that explained in Section 6.3, and follows the flow diagram shown in Figure 6.25. Similar to the planar design, first of all the lower cut-off frequency must be specified, which is key to determining the radiator height. This methodology assumes that the substrate has been previously chosen, and that the radiator will be fed by a coplanar waveguide (CPW), which has already been designed, based on classical microstrip impedance equations. [2] Then, only the radiator design on the substrate is

---

[2]By using CST Microwave Studio, it is possible to determine the CPW dimensions by using the macro *Calculate analytical line impedance*, where there is the option *Coplanar waveguide with ground*.

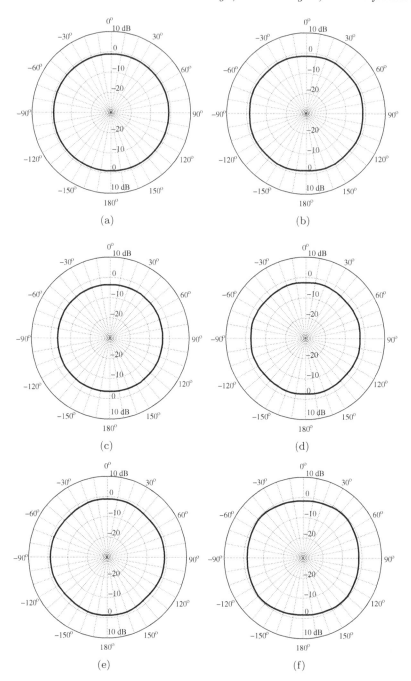

FIGURE 6.23
Radiation pattern for the antenna of Figure 6.21 at (a) 3 GHz, (b) 5 GHz, (c) 7 GHz, (d) 9 GHz, (e) 10 GHz and (f) 11 GHz.

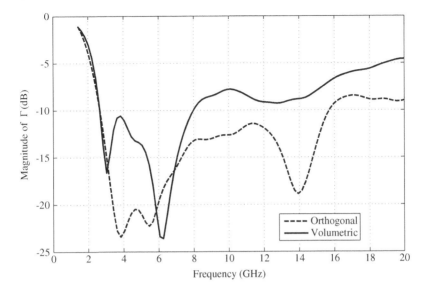

FIGURE 6.24
Comparison of the reflection coefficient magnitude of an orthogonal radiator and a revolution-solid (volumetric) radiator.

addressed. It also assumes the use of the symmetrical beveling technique as a mechanism to improve the impedance matching response.

As can be seen in Figure 6.25, it is again necessary to evaluate $\Gamma(x)$ but only for the bevel angle ψ and the radiator width W, which can be carried out following the general flow diagram seen in Figure 6.7.

Finally, it is worth noting that in order to increase the impedance bandwidth for a given bevel angle of a planarized antenna, an additional matching technique is necessary [5]. A possible way of achieving the above is by introducing a bevel in the ground plane around the CPW, both in the frontal part and the rear part of the substrate. It is determined by a simulation study that the ground plane bevel must start approximately 1/5 from the initial dimension of the radiator height. The radiator bevel is used to determine the bevel depth on the frontal part (let's say h_g) or on the rear part (let's say h_{gp}) of the substrate, as well as the distance between the radiator and the ground plane h; the expressions that relate these dimensions (depicted in Figure 6.26.) were experimentally determined by [5] and are given as:

$$h_g = -3.5 \tan \psi \qquad (6.1)$$

$$h_{gp} = -0.1 - 3.5 \tan \psi \qquad (6.2)$$

$$h = 0.7 - 3.5 \tan \psi \qquad (6.3)$$

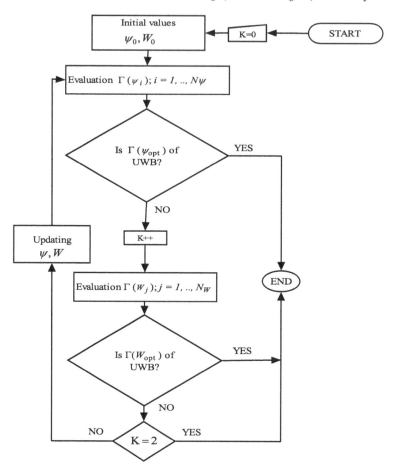

FIGURE 6.25

Flow diagram of the design methodology for planarized UWB omnidirectional antennas.

6.6.2   Preliminary square radiator

Once we have stated the general concepts described above, let us present the basic planarized structure. Thus, by assuming an arbitrary lower cut-off frequency of 2.36 GHz, we find that the radiator height is $l = 26.3\,\text{mm}$ according to Equation (4.17). This dimension is initially applied also to the radiator width (i.e., a square shape radiator). By calculating the CPW for an input impedance of $50\,\Omega$, the dimensions depicted in Figure 6.27 result. Since a height of the ground plane no longer of 10 times the width of the CPW is suggested [5], the value used here is 17 mm. As regards the substrate width, it is suggested to consider at least 10 mm additional to each radiator side [5], so the total ground plane width is 46.3 mm.

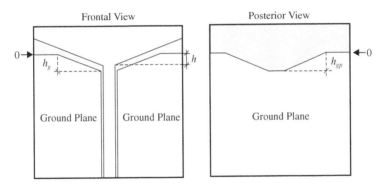

FIGURE 6.26
Geometry of the rectangular planarized UWB monopole with the bevel technique on the ground plane.

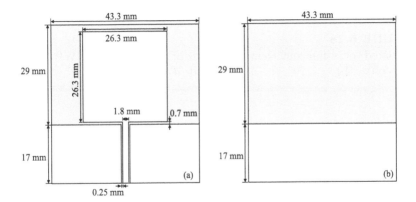

FIGURE 6.27
Preliminary geometry for a square planarized antenna: (a) Frontal view, (b) rear view.

6.7 Tuning of a Planarized Rectangular Antenna for UWB

6.7.1 Simulation results

Let us now start the design methodology described in Section 6.6. Consider the initial radiator illustrated in Figure 6.27 with an RT5880 substrate of 1.27 mm thickness ($\varepsilon_r = 2.2$), over which a bevel technique is symmetrically applied on its base according to the first evaluation step of the flow diagram in Figure 6.25. Figure 6.28 shows the simulated reflection coefficient magnitude for

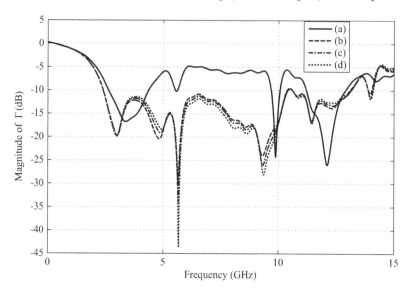

FIGURE 6.28
Simulated reflection coefficient magnitude of the planarized UWB antenna for
(a) $\psi = 0°$, (b) $\psi = 20°$, (c) $\psi = 22°$ and (d) $\psi = 24°$.

$\psi = 0°, 20°, 22°, 24°$ (although other values of ψ were also simulated, lower
bevel angles do not contribute to impedance matching improvement at all, so
they are not presented here). From these results, it is clear that the larger
the angle ψ, the better the impedance matching (represented by the down-
ward displacement of curves from the threshold of $|\Gamma| = -10\,\text{dB}$), where the
bevel angle $\psi = 22°$ provides the best conditions, because it has a median of
$|\Gamma| = -12.46\,\text{dB}$ for a percentile range of 74.2%. Although this geometry could
fulfill certain UWB expectations, a mismatching is found around 11 GHz.

The second step of the design methodology consists of varying the radiator width, such that a rectangular shape is obtained and a better impedance
matching condition is attained. Thus, the simulated reflection coefficient magnitude is shown in Figure 6.29, where the radiator width takes three values
$W = 30.3\,\text{mm}$, $W = 32.3\,\text{mm}$ and $W = 34.3\,\text{mm}$. By carefully analyzing
the results of $|\Gamma(W)|$, we observe that the condition of $W = 30.3\,\text{mm}$ intro-
duces a critical behavior around 11 GHz, where $|\Gamma| \approx -10\,\text{dB}$, whereas for
$W = 34.3\,\text{mm}$ the curve is near an impedance mismatching at 4 GHz. The
value of $W = 32.3\,\text{mm}$ conserves an acceptable impedance matching response
in the whole bandwidth from 2.24 to 12.62 GHz.

In terms of the radiation pattern, Figure 6.30 shows the simulated results
for the planarized antenna tuned with $\psi = 22°$ and $W = 32.3\,\text{mm}$ for 3, 5, 7,
9, 10 and 11 GHz. A quasi-omnidirectional radiation pattern can be observed
from this figure for almost all frequencies, except for 10 and 11 GHz.

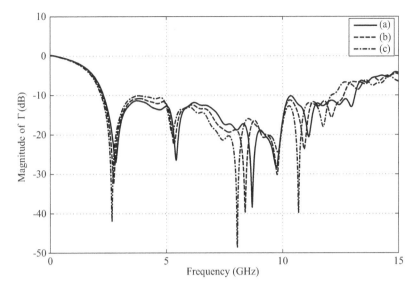

FIGURE 6.29
Simulated reflection coefficient magnitude of the planarized UWB antenna for three radiator widths: (a) $W = 30.3\,\text{mm}$, (b) $W = 32.3\,\text{mm}$ and (c) $W = 34.3\,\text{mm}$.

6.7.2 Measurement results

Let us now present the measurement results of the UWB planarized structure designed above. Figure 6.31 shows the prototype of this antenna, which was formed on a printed circuit board RT5880 with $\varepsilon_r = 2.2$ and 1.27 mm thick [10]. The reflection coefficient magnitude was measured in an Agilent E8362B network analyzer, results of which are shown in Figure 6.32. In this figure, a bandwidth reduction can be seen in comparison with the simulated results (see plot for $W = 32.3\,\text{mm}$ in Figure 6.29). Essentially, the lower cut-off frequency is shifted to around 3.5 GHz. Two critical impedance matching points can be seen now at 7.29 GHz and 9.69 GHz, where $|\Gamma| = -10.07\,\text{dB}$ and $|\Gamma| = -8.88$ dB, respectively. These critical frequencies (which are not present in the simulated results) are attributed to imperfections in the construction process.

The results corresponding to the radiation pattern at 3, 7 and 10 GHz are shown in Figures 6.33, 6.34 and 6.35, respectively, where the simulated radiation pattern is also included for comparison purposes. As can be seen, there is a high similarity between the graphs at 3 GHz. However, as the frequency is increased, significant differences emerge, which are associated with the presence of parasitic currents related to some errors in the antenna construction. In fact, this is in accordance with the impedance mismatching starting at the critical frequency of 7.29 GHz discussed in the previous paragraph.

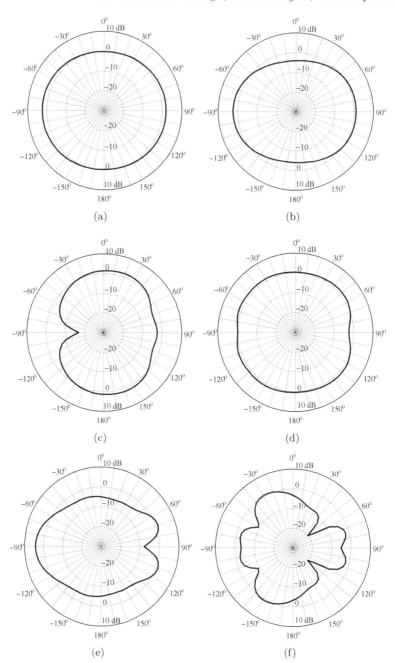

FIGURE 6.30
Simulated radiation pattern of the rectangular planarized antenna with $\psi =$ 22° and $W = 32.3$ mm at (a) 3 GHz, (b) 5 GHz, (c) 7 GHz, (d) 9 GHz, (e) 10 GHz and (f) 11 GHz.

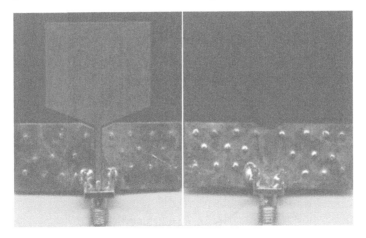

FIGURE 6.31
Prototype of a rectangular planarized antenna.

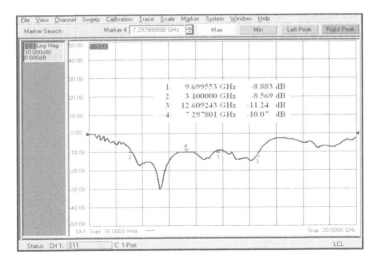

FIGURE 6.32
Measured reflection coefficient magnitude of the rectangular planarized antenna with $\psi = 22°$ and $W = 32.3$ mm.

6.8 Scaling Method to Achieve Other Bandwidths

Finally, it is worth explaining what is called the *scaling method*, which is a mechanism to define other frequency ranges where an antenna with new dimensions can operate. This method has been implemented on both planar

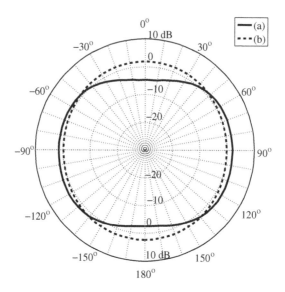

FIGURE 6.33
Radiation pattern at 3 GHz for the rectangular planarized antenna: (a) Measurement, (b) simulation.

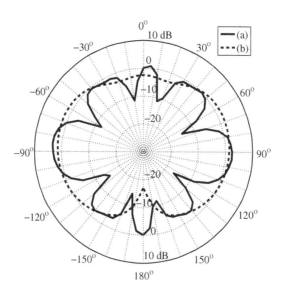

FIGURE 6.34
Radiation pattern at 7 GHz for the rectangular planarized antenna: (a) Measurement, (b) simulation.

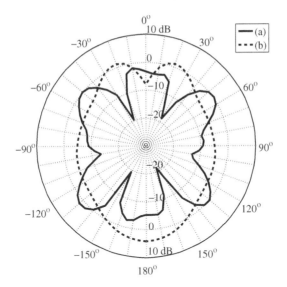

FIGURE 6.35
Radiation pattern at 10 GHz for the rectangular planarized antenna: (a) Measurement, (b) simulation.

and planarized UWB radiators by Peyrot-Solis [5] and Peyrot-Solis et al. [11]. Basically, the idea is centered on scaling the dimensions of a certain design to shift the frequency span for specific requirements. In the case of the RPMA tuned in Section 6.4, both dimensions ψ and W can be scaled, but variables a and h cannot be modified because they are related to the dimensions of its connector. Thus, with a well designed antenna, it is possible to vary the lower cut-off frequency, avoiding the design methodologies described in Sections 6.3 or 6.6. Then, a scaling factor (SF) is needed to uniformly modify the antenna dimensions. By considering the height l of the original radiator as a reference dimension and a new antenna height, l_s (remember that this variable is initially determined from the required lower cut-off frequency), the scaling factor can be directly derived as:

$$SF = \frac{l_s}{l} \tag{6.4}$$

For example, let us then scale the RPMA proposed in [5] in order to design a new UWB antenna for operation in another frequency range (let us say $f_L = 500$ MHz), and whose dimensions are depicted in Figure 6.36. Let us also take a margin of 40 MHz for any possible construction error. Then, by applying Equation (4.17) for a lower cut-off frequency of 0.46 GHz, a new radiator height of $l_s = 134$ mm is determined, and by substituting this value into (6.4), a factor $SF = 3.61$ is obtained.

By applying the scaling factor to the dimensions of Figure 6.36, a new, scaled antenna is configured, whose dimensions are depicted in Figure 6.37. It

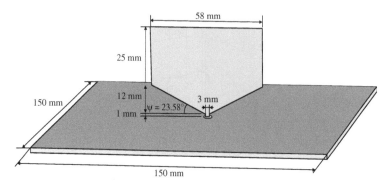

FIGURE 6.36
Geometry of an RPMA.

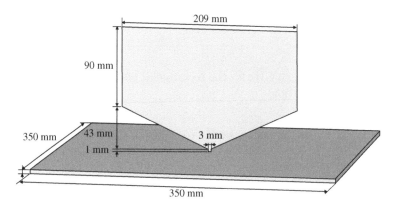

FIGURE 6.37
Geometry of a scaled RPMA.

is worth noting that, as previously pointed out, dimensions a and h were not modified and the ground plane size is now 350×350 mm^2 since it should be at least 2.5 times the radiator height [5].

Figure 6.38 shows the simulated reflection coefficient magnitude of the scaled antenna of Figure 6.37, where a shift in the lower cut-off frequency can be identified as $f_L = 660$ MHz (in the original design $f_L = 1.75$ GHz [5]). Although a difference of 200 MHz results with respect to the initial frequency ($f_L = 460$ MHz), the simplicity of the design method in using a scaling factor is attractive.

In order to conclude this section, let us examine the radiation pattern conditions for this scaled antenna, which were determined both by simulations and by measurements, and whose results are shown in Figures 6.39 and 6.40

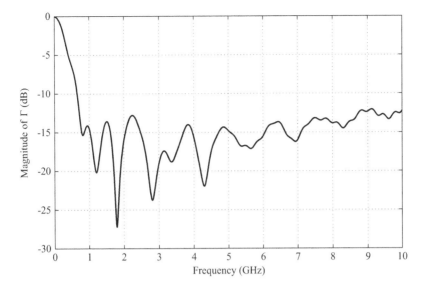

FIGURE 6.38
Simulated reflection coefficient magnitude of the scaled RPMA.

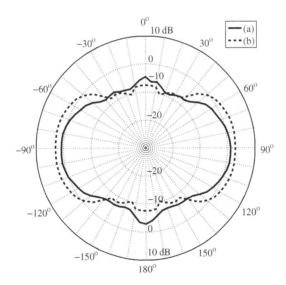

FIGURE 6.39
Radiation pattern at 5 GHz for the scaled rectangular planarized antenna: (a) Measurement, (b) simulation.

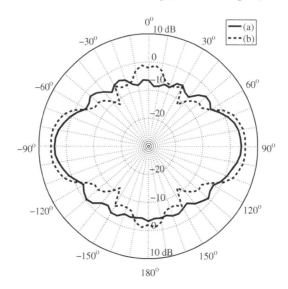

FIGURE 6.40
Radiation pattern at 10 GHz for the scaled rectangular planarized antenna:
(a) Measurement, (b) simulation.

for 5 GHz and 10 GHz, respectively. As can be seen, both graphs converge at
least in shape, although a difference in gain of around 1 dB is observed at two
frequencies.

Bibliography

[1] IEEE standard definitions of terms for antennas, IEEE std 145-1993,
1993.

[2] M. Hammoud, P. Poey, and F. Colomel. Matching the input impedance
of a broadband disc monopole. *Electronics Letters*, 29:406–407, 1993.

[3] S. W. Su, K. L. Wong, and C. L. Tang. Ultra-wideband square planar
monopole antenna for IEEE 802.16a operation in the 2–11 GHz band.
Microwave and Optical Technology Letters, 42(6):463–465, 2004.

[4] D. Valderas, I. Cendoya, R. Berenguer, and I. Sancho. A method of
optimize the bandwidth of UWB planar monopole antennas. *Microwave
and Optical Technology Letters*, 48(1):155–159, 2006.

[5] M. A. Peyrot-Solis. *Investigación y Desarrollo de Antenas de Banda Ultra Ancha (in Spanish)*. PhD thesis, Center for Research and Advanced Studies of IPN, Department of Electrical Engineering, Communications Section, Mexico, 2009.

[6] M. A. Peyrot-Solis, G. M. Galvan-Tejada, and H. Jardón-Aguilar. A $\pi/4$ bi-orthogonal monopole antenna for operation on and beyond of the UWB band. *International Journal of RF and Microwave Computer-Aided Engineering*, 21(1):106–114, 2011.

[7] M. J. Ammann. Control of impedance bandwidth of wideband planar monopole antennas using a beveling technique. *Microwave and Optical Technology Letters*, 30(4):229–232, 2001.

[8] M. A. Peyrot-Solis, G. M. Galvan-Tejada, and H. Jardón-Aguilar. Orthogonal ultra-wideband planar monopole antenna for EMC studies. In *VII International Sumposium on Electromagnetic Compatibility and Electromagnetic Ecology*, pages 141–144, 2007.

[9] M. Yanagi, S. Kurashima, T. Arita, and T. Kobayashi. A planar UWB monopole antenna formed on a printed circuit board. Technical report, Fujitsu Company, 2004.

[10] M. A. Peyrot-Solis, G. M. Galvan-Tejada, and H. Jardón-Aguilar. A novel planar UWB monopole antenna formed on a printed circuit board. *Microwave and Optical Technology Letters*, 48(5):933–935, 2006.

[11] M. A. Peyrot-Solis, J. A. Tirado-Mendez, G. M. Galvan-Tejada, and H. Jardón-Aguilar. Scaling factor in an ultra-wideband planar monopole antenna. *WSEAS Transactions on Circuits and Systems*, 8(5):1181–1184, 2006.

7

Design of Directional Planar and Volumetric UWB Antennas

CONTENTS

7.1   Introduction to Directional UWB Antennas

In this chapter, both planar and volumetric ultra wideband antennas are considered as possible configurations to achieve directional radiation characteristics in UWB antennas. Like the study reported in Chapter 6, not only are the impedance matching and phase response analyzed here, but also the radiation pattern behavior is considered. Nevertheless, provided that the considered designs have a directional radiation, different parameters should be taken into account: front to back lobe ratio (FBR), beamwidth at 3 dB or half power beamwidth (HPBW), side lobes character, and the presence of parasitic lobes.

Here, a different approach is taken from that in Chapter 6, in which a design methodology was initially introduced and the design, simulation and tuning of rectangular monopole antennas were carried out; instead, simulation and tuning of reported directional UWB antennas are presented first in this chapter. In this way, the early sections are dedicated to the analysis by computer simulation, using the CST Microwave Studio, of the performance of the following antennas:

1. Vivaldi Antenna: The Vivaldi antenna is a planarized radiator which possesses a theoretically infinite bandwidth, as explained in Chapter 3.

2. Leaf-shaped directional planar monopole antenna: The development of this antenna is based on the use of thicker substrates with a low dielectric constant. This antenna has three elements that increase the bandwidth: the tilt of the radiator which forms a certain angle with the ground plane, the substrate characteristics, and the leaf-shaped radiator.

3. Transverse electromagnetic (TEM) horn antenna: The structure of this radiator makes it of the volumetric type. Like the Vivaldi antenna, the TEM horn antenna presents excellent characteristics in the field of UWB antennas. This is the only antenna used as a standard to characterize other radiators.

4. Quasi-Yagi Antenna: The main advantage of this design is its reduced size in comparison to the previous antennas, but its main limitations are narrower bandwidth, and that it is a dispersive antenna.

After studying these antennas, some other proposals are analyzed, taking as a base a planar rectangular structure [1]. The design methodology explained in Chapter 6 is followed for its implementation, but now looking to achieve directional radiation features. In this way, a first directive design of a rectangular shape is achieved. Some modified versions of this single rectangular monopole antenna are also described. In particular, the inclusion of a reflector is considered as a possible alternative to increase antenna directivity. Other design aspects include extending the ground plane in such a way that a new ground plane structure is formed by two orthogonal surfaces. One of these surfaces has the function of a reflector.

Additionally, a design derived from the application of the planar-solid correspondence principle is also exposed, based on a recent work published in [2].

Basically a volumetric cone is transformed in a planar structure which provides a bandwidth larger than 10 GHz for a reflection coefficient magnitude lower than −10 dB and directive radiation characteristics. The evolution of the design of this antenna is also presented and its results are analyzed.

Finally, a comparison of some of the characteristics of the UWB directive antennas studied is presented, and performances are discussed, highlighting the advantages of each proposal.

7.2 Vivaldi Antenna

The Vivaldi antenna, as described in Chapter 3, is the planarized version of a horn antenna, offering a moderate gain in a low cost simple structure [3]. Originally, the Vivaldi antenna was designed as a balanced device, but currently there is an unbalanced version which eliminates the need for a UWB balun, which is an element that is very difficult to build.

Both versions of the antenna have been the subject of research, mainly in military applications due to their directional radiation pattern and gain of close to 10 dBi [4,5]. The antenna feeding, in general, determines the upper cut-off frequency [6], but theoretically it possesses an infinite bandwidth. On the other hand, the size of the slot determines the lower cut-off frequency.

The analysis procedure considered the characteristics established in [7]. Figure 7.1 shows the geometry and dimensions of the Vivaldi antenna sim-

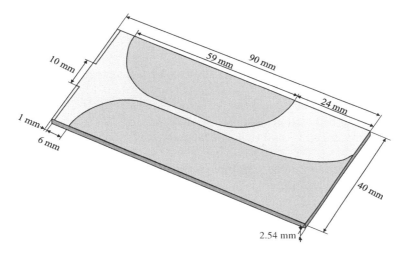

FIGURE 7.1
Geometry of the Vivaldi antenna.

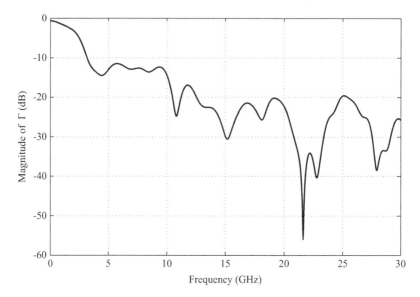

FIGURE 7.2
Simulated reflection coefficient magnitude for the Vivaldi antenna.

ulated in the CST Microwave Studio. The substrate employed in the design is an RT5880 Duroid, with thickness of 1.27 mm and a dielectric constant $\varepsilon_r = 2.2$.

The simulation results for both the magnitude and for the phase of the reflection coefficient are shown in Figures 7.2 and 7.3, respectively. As can be seen, the antenna has a bandwidth starting at 3.0 GHz and finishing at least at 30 GHz.[1]

In terms of the radiation pattern, the Vivaldi antenna shows a quasi-omnidirectional character below 5 GHz, as can be seen in Figure 7.4. However, as the frequency is increased, the radiation pattern tends to become more directive as can be noted in Figures 7.5, 7.6, 7.7 and 7.8. Therefore, in terms of directivity, it can be said that this antenna works properly from 5 GHz and beyond.

Let us now analyze the variation of some parameters of the radiation pattern as a function of the frequency. According to results of Figures 7.5, 7.6, 7.7 and 7.8, the gain varies from 3.5 to 5.4 dBi in the range of 5 − 8 GHz. From 9 to 16 GHz, the back and side lobes change, but the main lobe is almost conserved (the gain varies from 3.6 to 5.3 dBi). Between 17 GHz and 20 GHz, no substantial changes are observed in the main lobe, which shows gain values from 5.5 to 7.5 dB. The parasitic lobes in this frequency range are also

---

[1]This "upper" cut-off frequency obtained by simulation is limited by computational restrictions

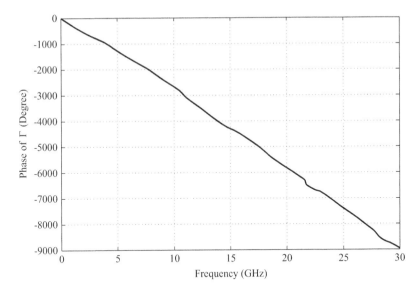

FIGURE 7.3
Simulated reflection coefficient phase for the Vivaldi antenna.

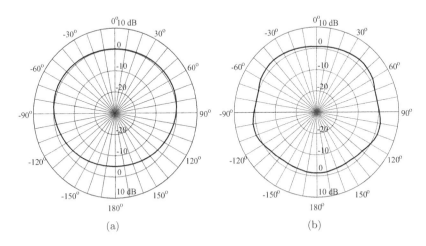

FIGURE 7.4
Simulated radiation pattern for the Vivaldi antenna at (a) 3 and (b) 4 GHz.

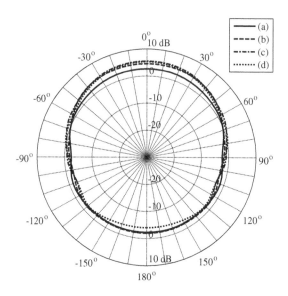

FIGURE 7.5
Simulated radiation pattern for the Vivaldi antenna at (a) 5 GHz, (b) 6 GHz, (c) 7 GHz and (d) 8 GHz.

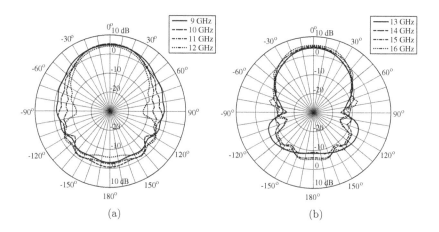

FIGURE 7.6
Simulated radiation pattern for the Vivaldi antenna at (a) 9–12 GHz, (b) 13–16 GHz.

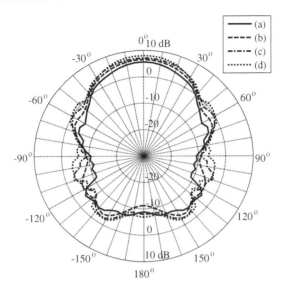

FIGURE 7.7
Simulated radiation pattern for the Vivaldi antenna at (a) 17 GHz, (b) 18 GHz, (c) 19 GHz and (d) 20 GHz.

stable compared to previous frequency intervals. Moreover, the back-to-front lobe ratio is smaller in comparison to the those of Figures 7.5, 7.6 reaching 15 dB at 17 GHz. Finally, for the $21-30$ GHz frequency range, the main lobe is preserved, and the gain goes from 8.4 to 9.3 dBi. In addition, the parasitic lobes are kept steady.

In order to evaluate the characteristics of the Vivaldi antenna designed in this text, its performance is compared to other Vivaldi designs. For example, let us first take as a comparison the design presented in [8], in which the same substrate, thickness and permittivity are used. The dimensions corresponding to the substrate width, $W_{substrate}$, and substrate length, $L_{substrate}$, are respectively calculated, in mm, as follows:

$$W_{substrate} = 0.68\lambda_L \tag{7.1}$$

$$L_{substrate} = 1.15\lambda_L \tag{7.2}$$

where λ_L is the wavelength (also in mm) at the lower cut-off frequency. Thus, for $f_L = 3$ GHz, $\lambda_L = 100$ mm and then $W_{substrate} = 68$ mm and $L_{substrate} = 115$ mm.

Another equation to determine the substrate width is introduced in [9]:

$$W = \frac{c}{2f_L\sqrt{\varepsilon_e}} \tag{7.3}$$

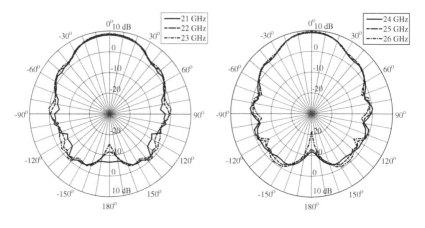

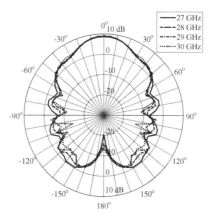

FIGURE 7.8
Simulated radiation pattern for the Vivaldi antenna at (a) 21–23 GHz, (b) 24–26 GHz and (c) 27–30 GHz.

with ε_e the effective permittivity, which is determined as $\varepsilon_e = \sqrt{\varepsilon_r - 1}$. Then, for the particular case at hand, where the lower cut-off frequency is 3 GHz and $\varepsilon_r = 2.2$, the substrate width is 45 mm.

Finally, another Vivaldi antenna is studied in [10], with dimensions of $100 \times 74 \text{ mm}^2$ and, although design equations are not presented, its lower cut-off frequency is reported as 2.9 GHz. Therefore, by comparing dimensions of Vivaldi antennas reported in [8–10], the resulting dimensions of the antenna presented in Figure 7.1 ($40 \times 90 \text{ mm}^2$) can be considered as satisfactory.

7.3 Leaf-Shape Antenna

The design of the leaf-shaped directional planar monopole antenna, as explained in Chapter 3, is based on the notion that the broadening of the bandwidth of a microstrip antenna requires the use of a thicker substrate, with low permittivity, such as air. This design shows three elements for improving the bandwidth: an inclined radiator forming a certain angle with the ground plane, a dielectric substrate, and a leaf-shaped radiator [11].

The minimum and maximum distances between the radiator and the ground plane are indicated in [11], with the former being 1 mm, and the latter approximately one half of the wavelength of the lower cut-off frequency. The antenna is tuned by varying the angle formed by the radiator and the ground plane. In [11], the optimum angle is set to 30°. A 50 Ω microstrip line is connected at the end of the radiator, which is parallel to the ground plane and centered to the radiator's symmetry axis.

In order to produce similar results to those of [11], a tuning process is carried out via simulations, which produces the geometry and dimensions illustrated in Figure 7.9. As can be seen, the "optimal" angle in this case is 40°.

The reflection coefficient magnitude for this antenna is shown in Figure 7.10. As can be seen, an improvement is generated since the bandwidth is wider, starting at 2.37 GHz and finishing at 30 GHz, compared to the antenna

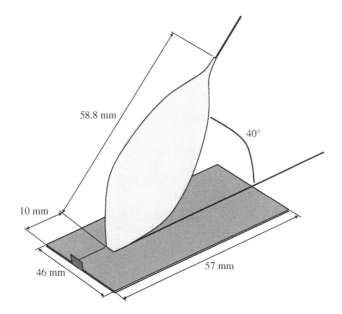

FIGURE 7.9
Geometry of the leaf-shaped directional planar antenna.

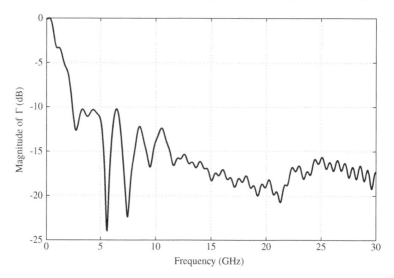

FIGURE 7.10
Simulated reflection coefficient magnitude for the leaf-shaped antenna.

in [11] where the limits are 3.05 GHz and 26.87 GHz, respectively. In comparison to the Vivaldi antenna, in this case, the phase of the reflection coefficient exhibits non-linear behavior (mainly for frequencies below 10 GHz), as shown in Figure 7.11.

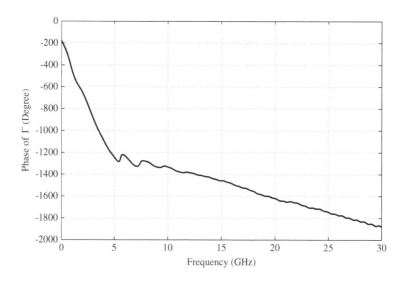

FIGURE 7.11
Simulated reflection coefficient phase for the leaf-shaped antenna.

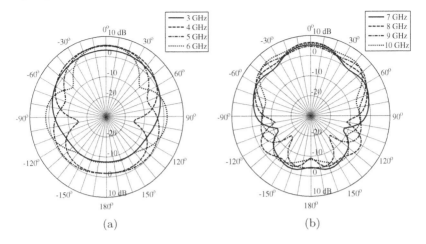

FIGURE 7.12
Radiation pattern for the leaf-shaped antenna at (a) 3–6 GHz and (b) 7–10 GHz.

Results concerning the radiation pattern of this antenna are divided in two groups. The first one ranges from 3 GHz to 10 GHz (see Figure 7.12), and the second one from 11 GHz to 30 GHz (see Figure 7.13). Unlike the Vivaldi antenna, the radiation pattern of the leaf-shaped antenna shows certain directivity from the lower cut-off frequency. Now, in the $3 - 10$ GHz interval, the main lobe has a gain of between 3.2 and 6.6 dBi and is almost unchanged, a behavior not observed for the parasitic lobes. For example, the radiation pattern at 6 GHz shows differences up to 5 dB at $\pm45°$, compared at different frequencies in this group where this effect is not present.

In the second frequency range, the gain of the antenna varies from 4.6 to 6.8 dBi. However, this parameter does not increase linearly as the frequency does. For example, the antenna has a gain of 4.6 dB at 26 GHz, while at 25 GHz, the gain is 5.1 dB. Moreover, at 11 GHz in the direction of $\pm60°$ there is a difference greater than 5 dB compared with the radiation pattern at other frequencies.

7.4 TEM Horn Antenna

The TEM horn antenna is, basically, an end-opened transmission line with plate shape [12, 13]. This type of antenna is generally formed of two sections of triangular-shaped metallic plates with a certain elevation angle. The height/width relation of the triangular-shaped plate keeps constant along the length of the antenna, with an aim to maintaining uniformity in the characteristic impedance.

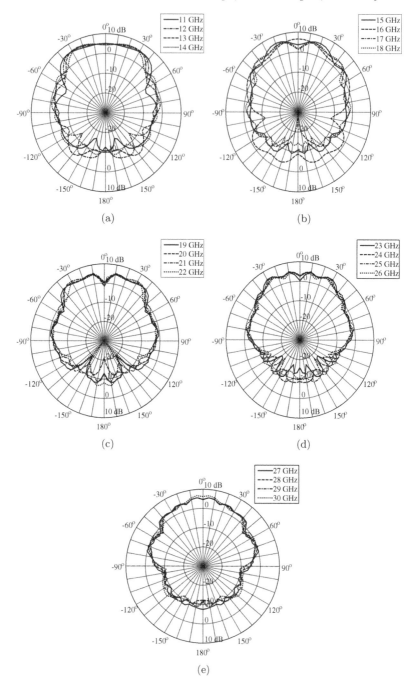

FIGURE 7.13

Radiation pattern for the leaf-shaped antenna at (a) 11–14 GHz, (b) 15–18 GHz, (c) 19–22 GHz, (d) 23–26 GHz and (e) 27–30 GHz.

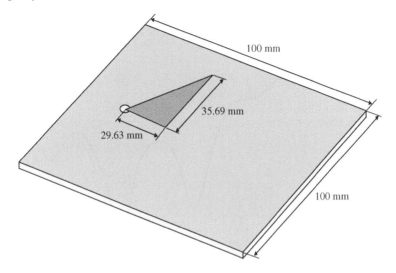

FIGURE 7.14
Geometry of a TEM horn antenna.

The UWB TEM horn antennas have moderate gain and relatively small dimensions. The transition between the feeding line and the antenna, as well as the antenna length and aperture, are important factors in determining the bandwidth [12, 14–16]. The upper cut-off frequency depends mainly on the aperture size, and also on the transition between the feeding line and the radiator. However, one of the limitations in terms of the bandwidth is that this quantity depends on the feeding structure. Thus, since this antenna is balanced, its bandwidth is limited by a balun. In order to change the operation from balanced to unbalanced, a ground plane is introduced instead of a triangular-shaped plate [17].

After a simulation process, the antenna geometry and dimensions shown in Figure 7.14 are defined for a TEM horn structure. The bandwidth achieved with this configuration is 11.84 GHz, with a lower cut-off frequency of 7.675 GHz and upper cut-off frequency of 19.52 GHz according to simulated results observed in Figure 7.15. In Figure 7.16 the reflection coefficient phase is depicted. The behavior of this parameter is linear up to 13 GHz, as can be observed.

The radiation pattern for different frequencies is presented in Figure 7.17. We can conclude from this figure that the radiation pattern shows some variations over the whole bandwidth, with a gain of between 1.5 and 8.1 dBi depending on the frequency increase. It is also important to note that the back and side lobes apparently behave steadily, except at 9, 16 and 19 GHz. At 9 GHz, for example, the pattern exhibits valleys whose amplitude is close to −23 dB toward 120°. At 16 GHz, the value is close to −18 dB pointing out

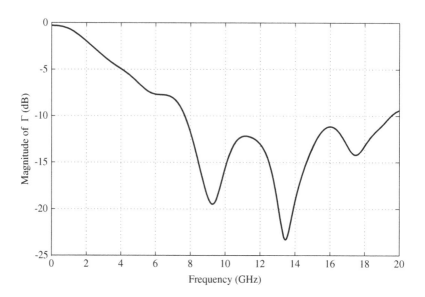

FIGURE 7.15
Simulated reflection coefficient magnitude for the TEM horn antenna.

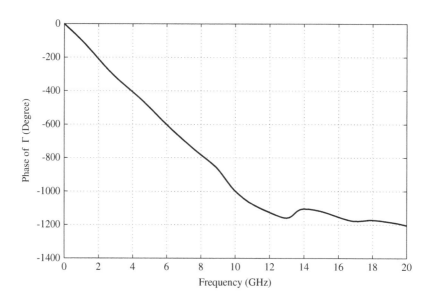

FIGURE 7.16
Simulated reflection coefficient phase for the TEM horn antenna.

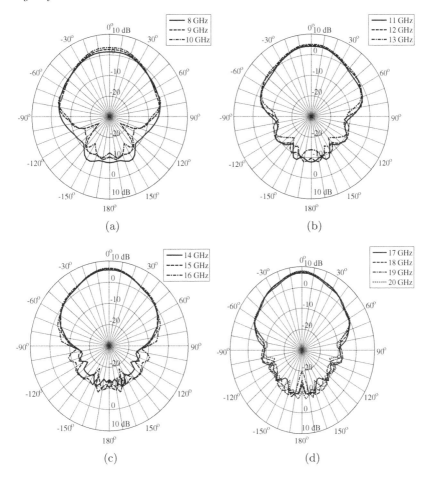

FIGURE 7.17
Radiation pattern for the TEM horn antenna: (a) 8–10 GHz, (b) 11–13 GHz, (c) 14–16 GHz and (d) 17–20 GHz.

150°. Finally, at 19 GHz, the null shows an attenuation of −20 dB at 180°. So, by analyzing the three UWB antennas simulated so far, it is the unbalanced TEM horn antenna that provides the most unchanged main lobe.

7.5 Quasi-Yagi Antenna

The quasi-Yagi antenna is a low cost directive antenna with moderate gain around 4.5 to 5.5 dBi [18]. Conventionally, the Yagi antenna is designed and built using wire dipoles or printed over a substrate. The classical wired

Yagi-Uda antenna is a device that has been used for many years, but it wasn't until 1991 that the version based on microstrip lines was proposed by Huang and Densmore [19].

Nowadays, planar Yagi antennas are considered to be the best alternative for communication systems, since they present some advantages over wire antennas, in that the substrate provides physical strength and a smooth transition with microstrip feeding lines. Thus, the selection of the substrate is a critical factor for the antenna performance. The substrate must have a high dielectric constant and a moderate thickness, depending on the application, since the antenna operation relies on the surface wave effect, which is in turn highly dependent on the material [18, 20].

The main disadvantage of the Yagi antenna is the need for a balun to match the balance mode of the radiator and the unbalance structure of the feeding line [21–23]. In [24] there is a proposal which employs a coplanar waveguide to overcome the problem caused by the microstrip line.

A specific design with good performance shows that the driver, directors, and ground plane are elliptical, which allows a reduction in size. This antenna is considered to be compact, since its dimensions, including the substrate effect, are $0.3\lambda_0 \times 0.5\lambda_0$, where λ_0 is the wavelength in free space at the central frequency [25]. The space between the microstrip lines barely affects the antenna bandwidth. Nevertheless, the most sensitive parameters are the driver length, and the space between this element and the reflector [20] and [26].

The design of this type of antenna starts by setting the driver length, which is almost $0.5\lambda_{eff}$, where λ_{eff} is the wavelength in the substrate at the effective lower cut-off frequency. According to the required parameters, the directors' length is around $0.45\lambda_{eff}$ [27]. The space between directors is around $0.1\lambda_{eff}$ to $0.2\lambda_{eff}$. Based on the the work of Kan et al. [25], a study of the performance of the antenna was made through a simulation process, whose geometry and dimensions are illustrated in Figure 7.18. The performance of the device was studied by using the CST Microwave Studio over an RT6010 Duroid substrate with relative permittivity $\varepsilon_r = 10.2$.

Compared to the Vivaldi and TEM horn antennas, the Yagi antenna has a narrower bandwidth. According to the results for the $|\Gamma|$ parameter shown in Figure 7.19, the bandwidth is almost 5.1 GHz, starting at 9.7 GHz and finishing at 14.8 GHz. The matching is deeper at 13.5 GHz, where $|\Gamma| = -21.6$ dB. The behavior of the reflection coefficient phase is non-linear, since this is a resonant antenna. Figure 7.20 shows the dependence of the reflection coefficient phase as a function of frequency. The most abrupt change of the phase slope occurs where the deepest matching is located.

The radiation pattern of the antenna is directional over the whole bandwidth. Figure 7.21 presents the behavior of the radiation pattern from 10 to 14 GHz. From this figure, we see that the main and back lobes are steady, although at 10 and 11 GHz a variation of 5 dB is found in the back lobe. The antenna gain shows values of 4.2 dBi at 14 GHz, and 6.7 dBi at 10 GHz. Also,

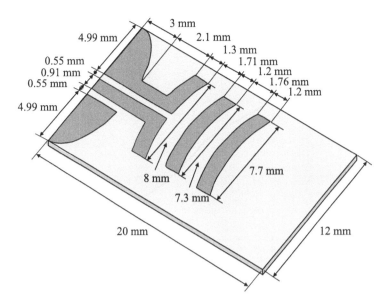

FIGURE 7.18
Geometry of the quasi-Yagi antenna.

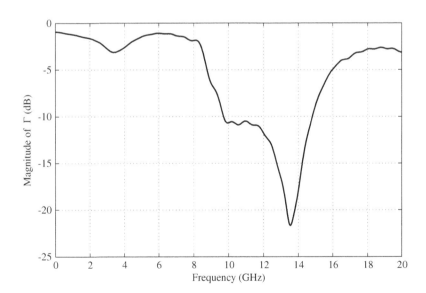

FIGURE 7.19
Simulated reflection coefficient magnitude of the quasi-Yagi antenna.

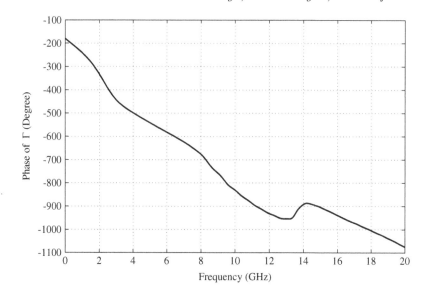

FIGURE 7.20
Simulated reflection coefficient phase of the quasi-Yagi antenna.

one can observe that the direction of the main lobe is perpendicular to the main axis of both the radiator and the director.

7.6　Design of a Directive Rectangular Planar Monopole

The idea of extending an omnidirectional rectangular PMA designed in Chapter 6, to a structure that produces a directional pattern, was explored in [1]. The development of this type of antenna is of great relevance, since the directivity of a UWB planar monopole antenna is one of the least studied parameters. In fact, to the knowledge of the authors, there is only one prototype of this kind in the literature, which was presented in [11].

Usually, designs of UWB directive antennas are based entirely on computer simulations, **where there is no methodology or definition of** equations to calculate the lower cut-off frequency. The existence of a procedure to develop this type of antenna promises to be very useful and of great interest, because directional radiators have a wide range of applications, such as electromagnetic spectrum monitoring, electromagnetic compatibility, radiofrequency weapons, and so on. The main results of this type of radiator were published in [1].

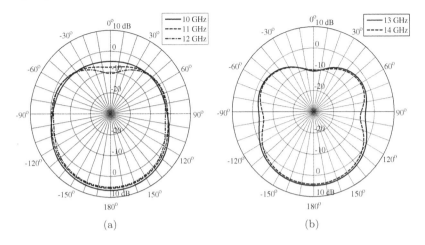

FIGURE 7.21
Simulated radiation pattern of the quasi-Yagi antenna: (a) 10–12 GHz and (b) 13–14 GHz.

7.6.1 Design methodology for a UWB directive rectangular PMA

A parameter critical to achieving a grade of directivity in a UWB antenna is the angle of inclination β formed between the radiator and the ground plane. Now, as explained in Section 4.8 (where β is taken from the normal to the radiator), this angle has an impact on the bandwidth due to the confinement of electromagnetic energy and, hence, it is important that it be considered for the antenna design. Essentially, the methodology presented in Section 6.3 is followed, but also including the evaluation of $\Gamma(\beta)$. Then, the procedure starts fixing the initial conditions of the radiator. In this case, the condition is that $\beta \neq 90°$, and the radiator height l is obtained by modifying Equation (4.17), such that,

$$f_L = \frac{62.1\,l}{l \sin \beta} \tag{7.4}$$

Initially setting $\beta = 30°$, which is the experimentally-determined minimum angle that gives an adequate impedance matching and f_L in GHz, the height of the radiator in mm is determined. This height is no longer modified along the design procedure. With this initial dimension, the ground plane is calculated, which must be at least twice as long as the height of the radiator, and 2.5 times its width. Up to this point, a rectangular radiator has been achieved, and now the methodology proposed in Figure 7.22 must be followed. Here, the radiator is perpendicular to the ground plane; therefore, during the early stages of the design, the lower cut-off frequency is below the required value.

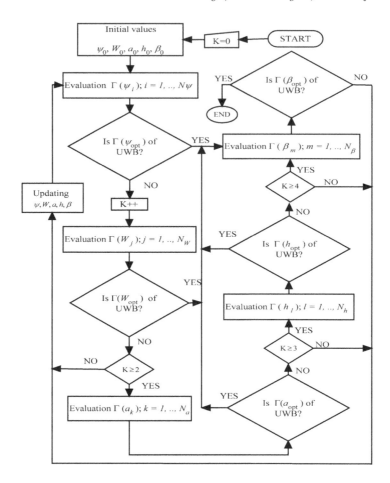

FIGURE 7.22
Flow diagram of the design methodology of a directive planar UWB antenna.

It is important to emphasize that the flow diagram in Figure 7.22 is very similar to that shown in Figure 6.6, but with the inclusion of an evaluation of the parameter Γ as a function of β. Another difference is that this evaluation is always carried out, although any other evaluation of the variables ψ, W, a or h satisfies the UWB conditions. The evaluation of $\Gamma(\beta)$ must always be accomplished for $m = 1,\ \dots\ N_\beta$, with N_β being the total number of increments of β. Like the evaluations of Γ as a function of ψ, W, a and h explained in Section 6.3, the evaluation of $\Gamma(\beta)$ can be conducted using the same general flow diagram of Figure 6.7, naturally for parameter β.

Once the initial antenna dimensions are obtained through Equation (7.4), the lower cut-off frequency must be verified. In [28] a theoretical equation to determine the lower cut-off frequency (let us say $f_{L_{omni}}$) for an omnidirectional RPMA is derived by applying the solid-planar principle to it (see Section 4.4):

$$f_{L_{omni}} \frac{288\pi l}{[W(2l-b) + a(b+2h) + 4\pi l^2]} \qquad (7.5)$$

with $f_{L_{omni}}$ in GHz and the antenna dimensions W, l, a, b and h in mm (see Figure 6.5). Although in the case of directional RPMA, the radiator has the same shape as its omnidirectional counterpart, the inclination angle is not considered in Equation (7.5). So, in a similar way, Equation (7.5) is modified to include the angle, β, such that,

$$f_{L_{dir}} = \frac{288\pi l}{[W(2l-b) + a(b+2h) + 4\pi l^2]\sin\beta} \qquad (7.6)$$

However, the lower cut-off frequency calculated by this equation can be changed if some type of reflector is employed to diminish the back side lobe. It is worth mentioning that after carrying out a computer simulation study of the effect of a reflector on the lower cut-off frequency, it was found that on average, this limit is shifted 0.5 GHz above the theoretical limit.

7.6.2 Example of design of the UWB directive rectangular PMA

Let us assume a lower cut-off frequency of 7 GHz for the directive antenna. First, assume that a certain margin is considered, such that $f_L = 6.55$ GHz. This margin is needed as the effect of the reflector, as well as inaccuracies in the construction process, must be taken into account. With this value, the height of the radiator is calculated from Equation (7.4), obtaining $l = 19$ mm. After obtaining the radiator height, the dimensions of the ground plane are found. As mentioned above, this must be at least $2.5\,l$ width, that is 47.5 mm. The length of the ground plane must be $2\,l$, which means 38 mm. Therefore, allowing for an uncertainty margin, let us consider the selected dimensions as 55 mm wide and 40 mm long. The position of the feeder on the ground plane is shifted from the center due to the inclination of the radiator. The position is determined by a displacement D:

$$D = \frac{l}{2}\cos\beta \qquad (7.7)$$

For this design, the displacement is equal to 8 mm. Figure 7.23 shows the geometry and initial dimensions of the first stage of the design. When the antenna is tuned by altering the variables ψ, W, a and h, according to the flow diagram in Figure 7.22, the next step is to tune the antenna by evaluating $\Gamma(\beta_m)$. For this last stage, assume that the antenna model in Figure 7.24 is employed, sweeping over five values of $5°$ of separation, corresponding to the range from $\beta = 20°$ to $\beta = 40°$.

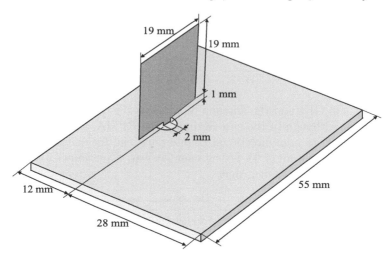

FIGURE 7.23
Initial geometry of the SPMA.

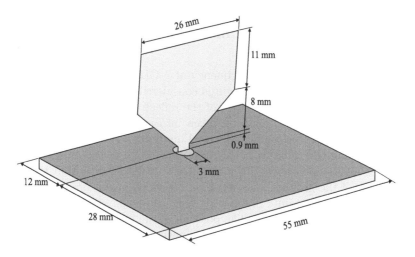

FIGURE 7.24
Geometry of the tuned directive RPMA.

After the simulation process, the results of the Γ parameter as a function of β depicted in Figures 7.25 and 7.26 are obtained for the magnitude and the phase responses respectively. As can be observed, abrupt changes are present among the different curves of $|\Gamma|$. This phenomenon may be due to the confinement of more electromagnetic energy when the radiator is inclined.

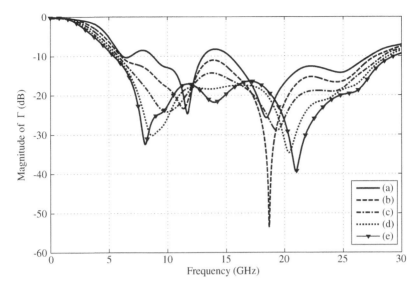

FIGURE 7.25
Simulated reflection coefficient magnitude of the RPMA for different tilt angles: (a) $\beta = 20°$, (b) $\beta = 25°$, (c) $\beta = 30°$, (d) $\beta = 35°$ and (e) $\beta = 40°$.

Therefore, as predicted in Chapter 4, the radiator tilt does not only affect the antenna directivity, but also the bandwidth.

Now, regarding this last parameter, from results shown in Figure 7.25, as long as the angle between the radiator and the ground plane is increased, a better impedance matching is attained. Moreover, this behavior is inversely proportional to the directivity, which implies a trade-off between directivity and bandwidth. Except for $\beta = 20°$, the reflection coefficient magnitude curves have an acceptable performance. The selected value is obtained by calculating the median and the percentile range which is $\beta = 30°$, achieving a bandwidth of 22.59 GHz (from 5.83 GHz to 28.42 GHz), with a median of $|\Gamma| = -16.18$ dB and percentile range of 77.7%. In terms of the phase performance, a quasi-linear behavior is observed for $\beta = 30°$ and $\beta = 35°$, whereas for the other tilt angles abrupt phase changes occur (see Figure 7.26). These changes are associated with the deep resonances shown in Figure 7.25.

From both Figures 7.25 and 7.26 it can be concluded that the angle β has an important role not only in radiation directivity (as will be discussed below), but also in impedance matching and phase characteristics. As already stated, the larger the value of β, the wider the bandwidth, thus the $|\Gamma|$ plots tend to move downwards. Nevertheless, the increase of β can introduce non-linearities in the phase of Γ, causing the pulse distortion problem. Hence, for the particular geometry and dimensions of the analyzed antenna, the best trade-off is for $\beta = 30°$.

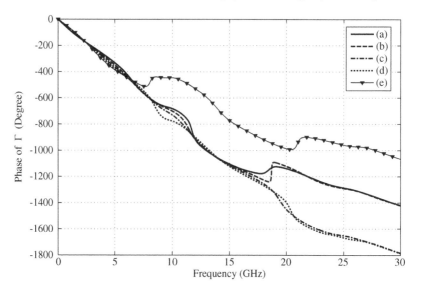

FIGURE 7.26
Simulated reflection coefficient phase of the RPMA for different tilt angles:
(a) $\beta = 20°$, (b) $\beta = 25°$, (c) $\beta = 30°$, (d) $\beta = 35°$ and (e) $\beta = 40°$.

Then, simulations are conducted in order to observe radiation pattern variations through the band of interest. Figure 7.27 shows the radiation pattern
at 6, 10, 14, 18, 22 and 26 GHz with $\beta = 30°$. As can be seen, the radiation
pattern becomes more directive as the frequency is increased.

7.6.3   Mechanical refinement of the antenna structure

Up to here, we have successfully designed a UWB directive planar monopole.
However, the radiator tilt used to confine the energy and focus the radiation in
a certain direction represents a mechanical limitation of the general structure.
After analyzing different options to increase the mechanic resistance, a support
on the radiator is considered to be a feasible alternative. To implement this
modification of the structure, a low dielectric constant and low loss tangent
material is chosen to avoid as far as possible any interaction with the antenna
parameters. So, the support is made as a teflon cylinder. It is important to
study the behavior of the antenna with this new part of the structure. The
geometry with the radiator support is shown in Figure 7.28.

Since a lower cut-off frequency of 6.5 GHz was initially defined and, by
simulation results, the prototype started to operate at 5.8 GHz, there is a
margin of 0.7 GHz. This margin allows the introduction of a post made of
teflon (see Figure 7.28) and a reflector which reduces the side lobes, keeping
the requirements of directivity. In order to evaluate the impact of the teflon

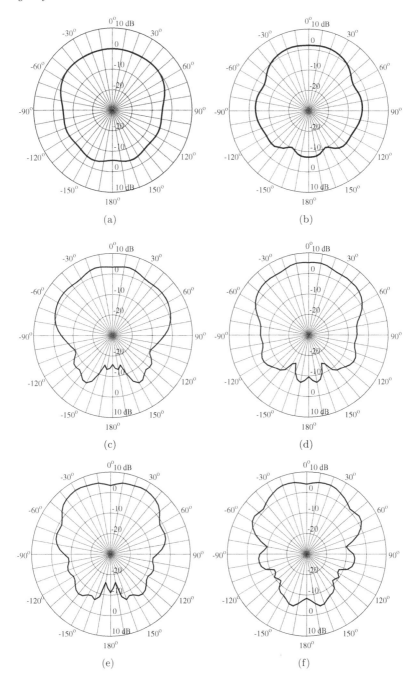

FIGURE 7.27
Simulated radiation pattern of the RPMA for $\beta = 30°$ at different frequencies: (a) 6 GHz, (b) 10 GHz, (c) 14 GHz, (d) 18 GHz, (e) 22 GHz and (f) 26 GHz.

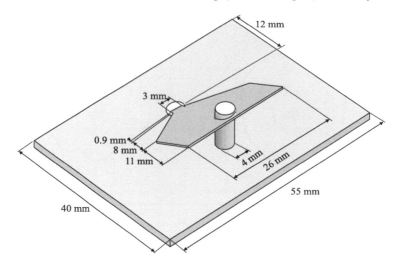

FIGURE 7.28
Geometry of the directive RPMA with a mechanical support.

cylinder, the magnitude of the reflection coefficient is analyzed from results of Figure 7.29, and its phase response from Figure 7.30. As seen in these figures, the magnitude and phase of Γ are not significantly modified.

Regarding the simulated radiation pattern, which is shown in Figure 7.31, it is concluded that from 6 to 10 GHz, no significant modifications are obtained when the teflon support is used, in comparison with results of Figure 7.27. But, at 14 GHz, a reduction of 4 dB of the back lobe levels is obtained, as well as an increase of gain at the main lobe close to 0.6 dB. Nonetheless, at 18 GHz an increase of 2 dB in the back lobe is observed, and at the same time, the main lobe is enhanced by 0.5 dB. At 22 GHz, the back lobe is slightly boosted, and the main lobe is increased by 0.8 dB. At 26 GHz no modifications are seen on either lobe.

7.6.4   Use of a reflector to increase the antenna directivity

In order to investigate a method of increasing the directivity of the antenna, a reflector is introduced as shown in [28,29]. This reflector not only modifies the radiation pattern, but also the bandwidth. So, to evaluate the effect caused by the reflector on these parameters, the main features of the antenna were studied through a simulation processes using the CST Microwave Studio [28,29]. According to their results, it is seen that the reflection coefficient magnitude is highly modified, since the resonances are reduced and the upper cut-off frequency is shifted to a higher frequency region. As a consequence, a wider bandwidth is attained. As regards the phase response, Figure 7.32 shows that this parameter maintains a quasi-linear behavior.

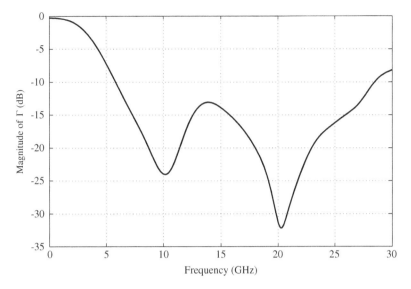

FIGURE 7.29
Simulated reflection coefficient magnitude of the directive RPMA with a mechanical support.

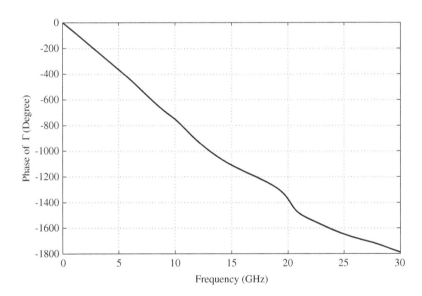

FIGURE 7.30
Simulated reflection coefficient phase of the directive RPMA with a mechanical support.

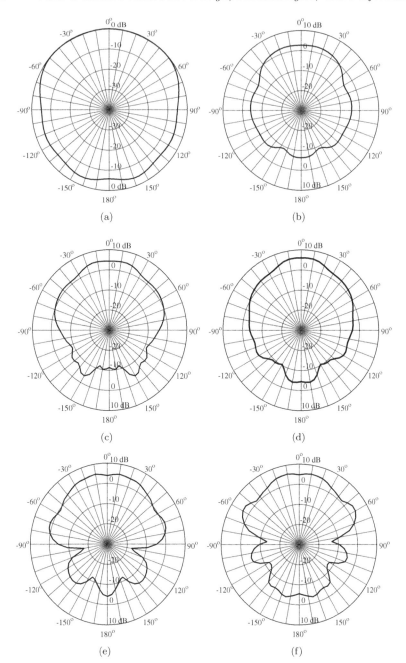

(a)

(b)

(c)

(d)

(e)

(f)

FIGURE 7.31
Simulated radiation pattern of the directive RPMA with a mechanical support. (a) 6 GHz, (b) 10 GHz, (c) 14 GHz, (d) 18 GHz, (e) 22 GHz and (f) 26 GHz.

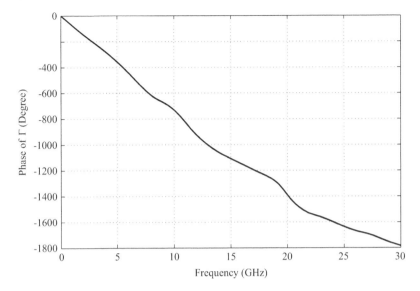

FIGURE 7.32
Phase response of the directive RPMA with reflector reported in [28].

The results of the radiation pattern obtained by computer simulation at 7, 10, 14, 18 and 26 GHz are presented in Figure 7.33. By comparing this pattern to those in Figures 7.27 and 7.31, certain differences can be observed: at 10 GHz there is a diminution of the back lobe close to 2.4 dBi; at 14 GHz the main lobe reaches 1.3 dB more, and there is a reduction of 10 dB of the back lobe; at 18 GHz the side lobe level is reduced 0.7 dB and 5 dB in the back lobe. Finally, at 26 GHz, the back lobe is 4 dB smaller. Therefore, the introduction of the reflector is an acceptable trade-off between the reduction of the back lobe, and the frequency shift of the lower cut-off frequency.

Table 7.1 shows some parameters determined from the simulated radiation pattern for a wide range of frequencies. We focus particularly on the gain, the front to back lobe relation (FBR), beamwidth at 3 dB (HPBW), and the main lobe direction. From these values an interesting result is observed: the main lobe presents fewer variations as a function of the frequency, and it has a moderate gain of 5.2 dBi on average. The back lobe is smaller compared to the corresponding results of the non-reflector antenna (see Figure 7.31) with an average level of 12.33 dB. Generally speaking, these characteristics are conserved in the frequency range of 7 to 22 GHz. Also note the beamwidth is relatively narrow (76°).

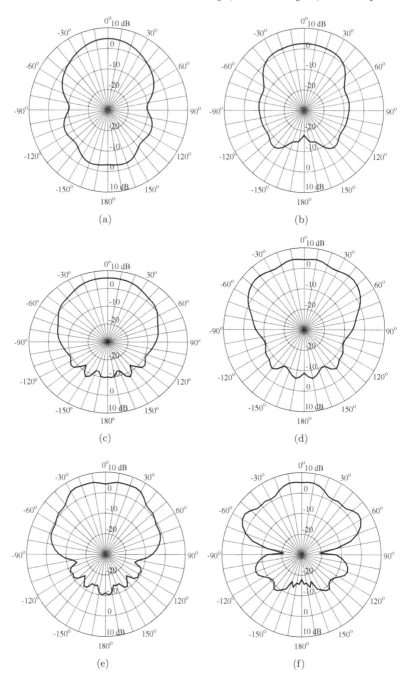

FIGURE 7.33
Radiation pattern of the directive RPMA with reflector reported in [28], (a)
7 GHz, (b) 10 GHz, (c) 14 GHz, (d) 18 GHz, (e) 22 GHz and (f) 26 GHz.

TABLE 7.1
Main features of the UWB directive RPMA at different
frequencies

Frequency (GHz)	Gain (dBi)	FBR (dB)	HPBW (degree)	Main lobe direction (degree)
7	4.7	7.5	61.1	0
8	3.1	9.9	78.4	0
9	1.9	10.0	90.3	0
10	2.5	9.0	85.7	0
11	4.8	8.7	69.8	0
12	5.6	8.4	65.4	10
13	5.4	12.0	64.3	10
14	6.0	13.5	67.1	0
15	5.9	12.7	90.6	10
16	5.9	12.7	90.2	15
17	5.4	12.0	101.7	10
18	4.7	10.4	123.2	10
19	6.3	13.3	40.4	355
20	7.1	14.1	42.6	0
21	6.4	15.0	63.6	10
22	5.8	15.7	74.1	25
23	5.8	13.4	90.0	25
24	4.8	14.8	135.0	25
25	4.4	13.7	135.9	60
26	5.9	14.9	41.7	355
27	6.6	15.6	42.9	10
28	6.3	15.3	52.5	15
29	5.5	14.5	59.5	15
30	3.6	8.9	55.0	20

7.6.5 Modification of the ground plane to improve the antenna directivity

As discussed in Section 4.6, it is possible to configure the antenna ground plane in a perpendicular structure similar to that in Figure 4.6. In essence, then, the ground plane is made up of two surfaces orthogonal to each other, where one is larger than the other, which are represented by their lengths, L_M for the major surface and L_m for the minor surface. This configuration allows a higher concentration of electromagnetic energy, so increasing its gain and reducing the back lobe level. Yao et al. suggest that the feed-point must be located in a lateral side in this type of structure [11].

Although the same design methodology for directive UWB planar antennas explained in Section 7.6 can be applied, we expect to require more iterations due to the new orthogonal ground plane. Like the single ground plane design, the electrical length l of the radiator must be determined. This dimension is useful for calculating the perpendicular ground plane dimensions.

Experimentally we found that both surfaces must have a width of at least $1.3\,l$, whereas for the main surface, it has to be 3 mm larger than $l\sin\beta$ and for the minor surface, it has to be 6 mm larger than $l\cos\beta$.

The main results of modifying the single ground plane into an orthogonal structure for the directive RPMA can be found in [29], where a higher gain, a larger bandwidth, and acceptable variations of the radiation pattern as a function of the frequency (at least by frequency groups) are reported for the dimensions obtained by the tuning process based on the design methodology of directive planar antennas explained here.

7.7   Design of Planar Directional UWB Antenna for Any Desired Operational Bandwidth

The problem of achieving a directional radiation pattern has been also recently studied in [2], whose design comes from a volumetric antenna to which the solid-planar correspondence principle is applied. As was explained in Chapter 4, early mechanisms to attain wider bandwidths were through volumetric structures (the classical example of these early designs being the biconical antenna). For the sake of simplicity, let us briefly define here the concept of this principle explained in Chapter 4 and describe its variables. Then, the solid-planar correspondence principle comes from the single cylindrical monopole with length l and radius r_d. When its radius is large, it can be seen as a volumetric structure over which the current is distributed. This element works as a comparison base for other planar structures, where their areas are made equal. In other words, this principle states that for any surface-revolution structure, there exists its counterpart planar antenna, so it is possible to get a planar structure version of any volumetric radiator [2]. In what follows, different aspects of how, based on a classical volumetric omnidirectional design, a directive UWB is accomplished, and its performance results are discussed. All this material is taken from [2].

7.7.1   Basic structure

The design process begins with a single conical antenna of height l, aperture angle α and diameter of the circular aperture d_c settled on a large flat ground plane with an input impedance of $50\,\Omega$. The well-known theory of the conical antennas [30] states that for $\alpha = 90°$ low variations on the reactive part of the antenna impedance over a very wide range of frequencies and radiator lengths can be achieved. On the other hand, $l = 0.24\lambda$ can be taken to fade the imaginary part of the input impedance to zero [27] and $d_c = 2l$. Let us assume that a resonant frequency of 2.4 GHz is desired; then the monopole length must be equal to $l = 30\,\text{mm}$ and the cone diameter of $d_c = 60\,\text{mm}$. The ground plane is approximately equal to λ. The geometry and dimensions of

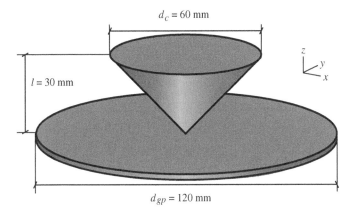

FIGURE 7.34
Conical antenna model with circular aperture [2].

this structure are shown in Figure 7.34. Simulations results (not shown here) indicate that the lower cut-off frequency of this initial antenna is 2.3 GHz with a bandwidth wider than 17 GHz and an omnidirectional radiation pattern.

7.7.2 Transformation process into directional radiation pattern

In order to initiate the directional process of the antenna design it is necessary, first of all, to change the circular aperture of the cone to an elliptical one with an eccentricity of, for our example, 0.44 but with the same ground plane. This change in the aperture maintains a similar reflection coefficient magnitude but with a slightly directional radiation pattern. After that, the radiator can be slanted [31], so the angle is relative to the vertical axis of the cone, as can be seen in Figure 7.35. By taking different slant angles (through variation of the displacement relative to the vertical axis of the cone), no major modification is presented in the reflection coefficient magnitude and therefore the operational bandwidth is not significantly affected. Nevertheless, the antenna gain goes from 2.5 to 7.9 dBi and the presence of secondary and back lobes is noted as the frequency is higher.

Finally, to reduce the back lobe at low frequencies and increase the gain of the volumetric antenna, a reflector was introduced through a tuning process to determine its dimensions and to reduce non desired effects to the impedance matching bandwidth. Thus, a size of 90 mm × 60 mm was determined with a position of the reflector with respect to the feeding point of 10 mm. Figures 7.36, 7.37 and 7.38 show, respectively, the antenna model with the reflector, the reflection coefficient magnitude, and the radiation pattern in the XY plane of this step of the design.

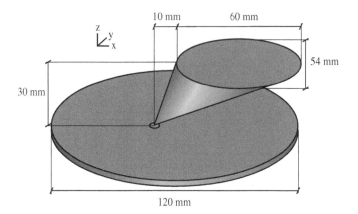

FIGURE 7.35
Conical antenna with elliptical aperture displaced 40 mm [2].

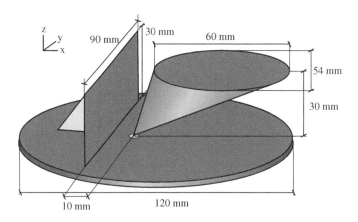

FIGURE 7.36
Conical antenna with elliptical aperture and reflector [2].

As can be seen, the effect of the reflector is translated into a reduction of the lower cut-off frequency from 2.4 to 1.8 GHz, a variation of the antenna gain from 5 to 7.9 dBi as frequency is higher, and a reduction of the back lobes (see Figures 7.37 and 7.38, respectively).

7.7.3   Application of the solid-planar correspondence principle

So far, a directional UWB antenna has been accomplished; however, it is of the volumetric type. The next step is then evolving this antenna to a planar

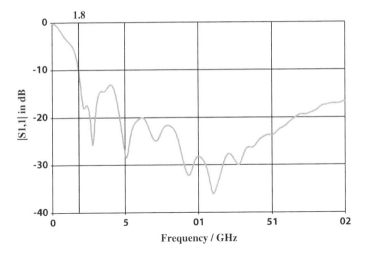

FIGURE 7.37
Reflection coefficient magnitude of the conical antenna with elliptical aperture and reflector [2].

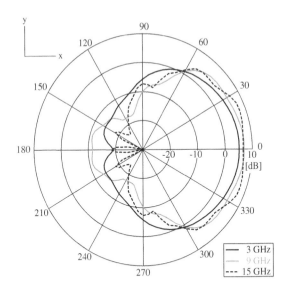

FIGURE 7.38
Radiation pattern of a conical antenna with elliptical aperture and reflector in the XY plane [2].

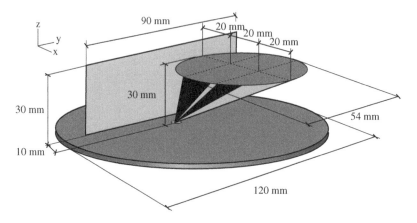

FIGURE 7.39
Semiplanar conical antenna with elliptical aperture and reflector [2].

one through the planar-solid correspondence principle applied in two axes. Although its elliptical aperture structure is preserved, a clear transformation can be appreciated in Figure 7.39, which is named *semiplanar conical antenna*. From these modifications, the lower cut-off frequency was shifted from 1.8 to 2.1 GHz, but the upper cut-off frequency is still larger than 20 GHz. On the other hand, the main lobe stability at high frequencies was affected.

Later on, the elliptical aperture structure of the antenna is removed to obtain a fully planar UWB antenna (and therefore it is called a *planar directional antenna*) and the geometrical distribution and dimensions are depicted in Figure 7.40. As can be seen, this antenna is formed by three triangular plates, two of them isosceles, and the other one scalene. The scalene triangle is at the desired main lobe direction, and the isosceles triangles are at the perpendicular position. The ground plane is circular with the feed point displaced from its center. The relationship $w = 1.66\, l$ was obtained through simulations.

For the geometry and dimensions given in Figure 7.40, simulations show that the impedance bandwidth and the gain were not affected substantially. The total effect of the evolution from volumetric to planar was an increase of 0.3 GHz in the lower cut-off frequency and the gain values also do not vary significantly in the simulated radiation pattern for all the frequencies considered (see Figure 7.41 and Table 7.2, respectively).

7.7.4   Design equation

Once the radiator is totally planar, a design equation for a certain desired lower cut-off frequency can be derived. As was explained in Chapter 4, the above is based on the theory that a planar monopole antenna can be seen as a

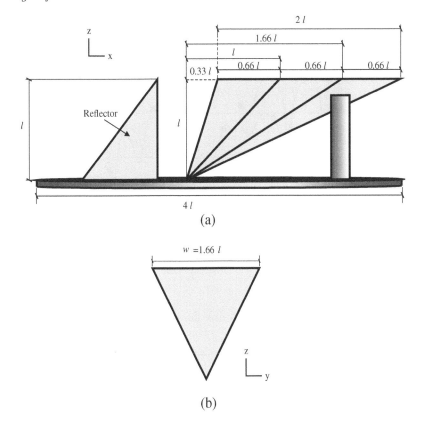

FIGURE 7.40
Geometrical distribution of the planar directional UWB antenna (a) Lateral view (b) front view of the isosceles triangle [2].

cylindrical monopole with a very wide effective diameter. Therefore, the lower cut-off frequency can be obtained through the equation to find the monopole length for the real input impedance given in Chapter 4 and rewritten here:

$$l = (0.24)\lambda F \tag{7.8}$$

where F is a term known as the length-ratio equivalent. As was already stated in Chapter 4, the term F is used to determine an equivalent area between a cylindrical monopole and a planar monopole radiator through the expression

$$F = \frac{l}{r_d + l} \tag{7.9}$$

with r_d the radius of the cylindrical monopole in mm. Then, for the structure

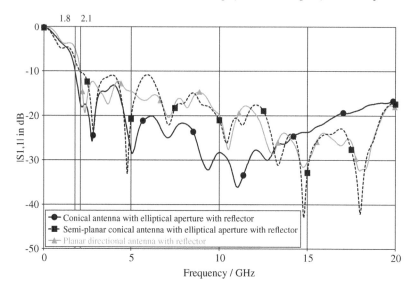

FIGURE 7.41
Reflection coefficient magnitude from volumetric to planar antenna evolution
[2].

presented in this section, the design procedure consists of making the radiator
area equal to the cylindrical monopole area (note that each triangle of the
planar UWB antenna represents $1/3$ of the total area of the antenna). In
other words,

$$2\pi r_d l = \frac{1}{3}\left(\frac{wl\sqrt{2}}{2}\right) + \frac{1}{3}\left(\frac{w\sqrt{l^2+w^2}}{2}\right) + \frac{1}{3}\left(\frac{lL}{2}\right) \qquad (7.10)$$

Provided that $L = 2\,l$ and $w = 1.66\,l$ and replacing in Equation (7.10), it
is possible to find that,

$$r_d = 0.2\,l \qquad (7.11)$$

Replacing (7.11) in (7.9) and in turn in (7.8), a relationship between the
radiator height and the desire lower cut-off frequency for the proposed planar
directional UWB antenna can be found as

$$f_L = \frac{60}{l} \qquad (7.12)$$

where the f_L represents the lower cut-off frequency in GHz and l in mm. As
can be seen, from Equation (7.12) it is possible to design the dimensions of
the antenna for any desired lower cut-off frequency. Then a validation of this
expression is necessary, which will be addressed in what follows.

TABLE 7.2

Comparison of some parameters of directional antennas [2]

Frequency (GHz)	Boresight Gain (dB)	3 dB beamwidth (degree)	Front-to-back ratio (dB)
Conical antenna with elliptical aperture and reflector			
3	5.1	98.3	-19.4
6	8.0	66.8	-14.9
9	7.0	103.6	-18.0
12	7.2	96.5	-20.2
15	7.8	73.9	-26.0
18	6.8	88.0	-18.0
Semiplanar conical antenna with elliptical aperture and reflector			
3	4.6	114.5	-19.5
6	8.5	50.7	-28.0
9	7.7	87.9	-15.5
12	6.6	90.5	-15.8
15	7.1	66.4	-17.0
18	5.8	85.0	-15.0
Planar directional antenna			
3	5.3	98.6	-14.9
6	7.6	65.4	-18.0
9	7.8	66.1	-16.5
12	8.3	42.7	-18.0
15	6.6	66.9	-16.5
18	6.4	70.0	-15.5

7.7.5 Experimental results

In order to validate Equation (7.12) two prototypes were built. The first prototype was designed for a lower cut-off frequency of 2 GHz and, therefore, by applying Equation (7.12) and all relations derived from it, dimensions given in Figure 7.42 were obtained. In a similar way, the second prototype was designed for a lower cut-off frequency of 3 GHz, whose resulting dimensions are depicted in Figure 7.43.

Both prototypes were evaluated using an Agilent NPA Series Network analyzer E8362B calibrated to a 50 Ω SMA connector. From the dimensions given for the first prototype simulated and measured lower cut-off frequency of 2.09 and 2.19 GHz were, respectively, obtained and a measured bandwidth of 16.2 GHz (see Figure 7.44). Regarding the second prototype, it presented a simulated and measured lower cut-off frequency of 2.91 and 2.8 GHz, respectively, and a measured bandwidth of 17 GHz (see Figure 7.45). By comparing all lower cut-off frequencies (desired, simulated, and measured), one can note that the dissimilarities among them are less than 10%, from which it can be concluded that the derived design equation provides a suitable form to determine the dimensions of new designs operating at different frequencies.

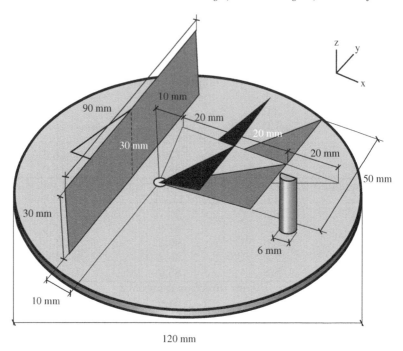

FIGURE 7.42
Planar directional UWB antenna for a lower cutoff frequency of 2 GHz [2].

Finally, it is worth presenting the results in terms of the radiation characteristics. The behavior of the simulated and measured radiation pattern for the first and second prototype can be seen in Figures 7.46 and 7.47, respectively. A good agreement occurs between the simulated and measured radiation patterns for both prototypes. In both cases, the same trend on the variation of the antenna gain through frequency can be observed.

7.8   Comparison of Different Directive UWB Antennas

With the different UWB directive antennas analyzed throughout this chapter, some of their key characteristics are summarized in Table 7.3 in order to have a basis for comparison (it is worth noting that in the case of the RPMA, only the single structure is considered). These antennas were designed considering initially the original dimensions in [7, 11, 14, 25, 28], which have been tuned here to obtain the best features. In order to make a fair comparison, these results are derived from a simulation process, since only the RPMA and the planar directive antenna exposed in Section 7.7 (PDA) were built.

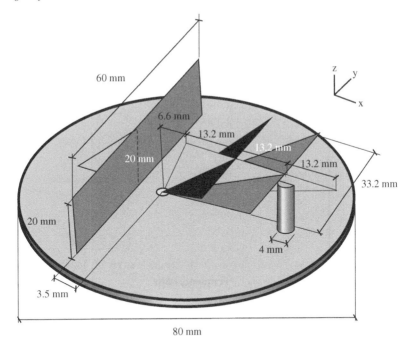

FIGURE 7.43

Dimensions of the planar directional UWB antenna for a lower cutoff frequency of 3 GHz [2].

So first of all, let us review the parameter often mentioned throughout this book: the bandwidth. Specifically, this quantity is related to the reflection coefficient magnitude (i.e. the impedance bandwidth). As can be appreciated, the Vivaldi antenna presents the widest bandwidth, although the leaf-shaped antenna shows a similar value, followed by the RPMA. Nevertheless, remember that the Vivaldi antenna was simulated up to 30 GHz due to computational restrictions, and consequently it is difficult to compare its real bandwidth (in principle wider than the simulated one). On the other hand, the quasi-Yagi antenna exhibits the narrowest bandwidth, which is not within the UWB band defined by the FCC provided that $f_L = 9.1$ GHz and $f_H = 14.8$ GHz. Naturally, its bandwidth fulfills the definition of a UWB bandwidth ($BW > 500$ MHz). In terms of the lower cut-off frequency, both the Vivaldi and the leaf-shaped antennas have a value below the limit of 3.1 GHz (although as was seen in Section 7.7 the PDA can be designed for any lower cut-off frequency). Regarding the upper cut-off frequency, in essence all designs surpass the FCC upper limit of 10.6 GHz. The fact that a device presents a bandwidth wider that than stated by the FCC can be attractive; however, an electromagnetic compatibility analysis must be carried out to

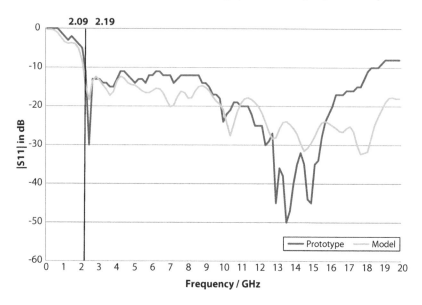

FIGURE 7.44
Simulated and measured reflection coefficient magnitude of the first prototype [2].

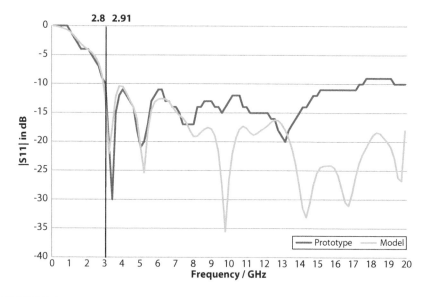

FIGURE 7.45
Simulated and measured reflection coefficient magnitude of the second prototype [2].

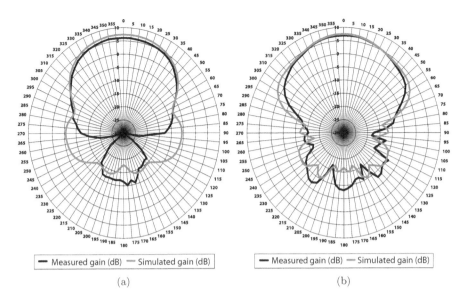

FIGURE 7.46
Simulated and measured radiation pattern at (a) 5 GHz and (b) 10 GHz of the first prototype [2].

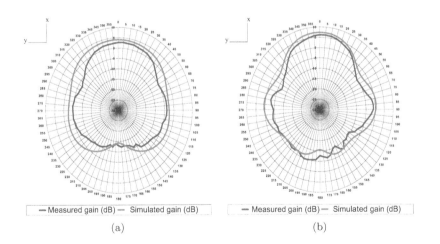

FIGURE 7.47
Simulated and measured radiation pattern at (a) 5 GHz and (b) 10 GHz of the second prototype [2].

TABLE 7.3

Comparison of directive UWB antennas

Characteristics	PDA[†]	RPMA	Vivaldi	Leaf-shaped	TEM horn	Quasi-Yagi
Ground plane dimensions (mm^2)	11,310	2200	3600	2622	10,000	240
Radiator dimensions (mm^2)	3419	392	1600	1560	529	65
Bandwidth (GHz)	> 17.9	23.4	30	27.63	11.84	5.1
f_L (GHz)	2.08	6.6	3.0	2.37	7.68	9.71
f_H (GHz)	> 20	30	30	30	19.52	14.8
Average gain (dBi)	6.99	5.18	6.21	5.62	5.4	5.32
Average FBR (dB)	17.16	12.33	10.94	12.56	13.54	12.24
Average HPBW (degree)	69.08	75.88	82.82	79.97	55.08	101.46
Design equations[‡]	Yes	Yes	Yes	No	Yes	Yes

[†] These values correspond to dimensions of the first prototype
[‡] Design equations are referred to determine the lower cut-off frequency

guarantee an adequate coexistence of wireless equipment sharing common frequencies.

By analyzing other characteristics given in Table 7.3, we found that the gain and back lobe level for all antennas are very similar in practical terms. The beamwidth at 3 dB is approximately of the same order for the PDA, RPMA, the Vivaldi, and the leaf-shaped antennas, whereas the quasi-Yagi has the widest value in spite of the array elements. In the case of the TEM horn antenna, its HPBW is the narrowest, which is expected since the horn structure allows its radiated electromagnetic energy to concentrate, which is the reason (as pointed out in Chapter 3), it is used as a standard antenna.

In terms of dimensions, the quasi-Yagi antenna presents the smallest values for both the substrate and the radiator, which is one of the reasons this device is attractive. The RPMA and the TEM horn antenna have similar radiator dimensions and they can be considered as moderate in comparison to those of the PDA, Vivaldi, and the leaf-shaped antennas.

Finally, but no less important, let us make some comments about the design equations. As marked at the bottom of Table 7.3, we are referring to equations to determine the lower cut-off frequency. In Chapter 4 the importance of this frequency in the UWB context was stressed, because there is a relationship between it and the antenna dimensions. Hence, the inclusion of these equations is fundamental for the global design of the antenna. Of all antennas discussed here, only the leaf-shaped structure does not provide this.

Bibliography

[1] M. A. Peyrot-Solis, G. M. Galvan-Tejada, and H. Jardón-Aguilar. Directional UWB planar antenna for operation in the 5–20 GHz band. In *17th International Zurich Symposium on Electromagnetic Compatibility*, pages 277–280, 2006.

[2] M. A. Peyrot-Solis, G. M. Galvan-Tejada, and H. Jardón-Aguilar. Proposal of a planar directional UWB antenna for any desired operational bandwidth. *International Journal of Antennas and Propagation*, 2014:1–12, 2014.

[3] H. Schantz. *The Art and Science of Ultra Wideband Antennas*. Artech House, Norwood, MA, 2005.

[4] E. Gazit. Improved design of the Vivaldi antenna. *IEE Proceedings*, 135(2):89–92, 1988.

[5] J. P. Weem, B. V. Notaros, and Z. Popovic. Broadband element array considerations for SKA. *Perspectives on Radio Astronomy Technologies for Large Antenna Arrays, Netherlands Foundation for Research in Astronomy*, pages 59–67, 1999.

[6] M. A. Peyrot-Solis, G. M. Galvan-Tejada, and H. Jardón-Aguliar. State of the art in ultra-wideband antennas. In *II International Conference on Electrical and Electronics Engineering (ICEEE)*, pages 101–105, 2005.

[7] The 2000 CAD benchmark unveiled. *Microwave Engineering Europe Magazine*, pages 55–56, 2000.

[8] S. G. Kim and K. Chang. Ultra wideband 8 to 40 GHz bran scanning phased array using antipodal exponentially-tapered slot antennas. In *2004 IEEE MTT-S International Microwave Symposium Digest*, pages 1757–1760, 2004.

[9] S. Wang, X. D. Chen, and C. G. Parini. Analysis of ultra wideband antipodal Vivaldi antenna design. In *Loughborough Antennas and Propagation Conference*, pages 129–132, 2007.

[10] K. V. Dotto, M. J. Yedlin, J. Y. Dauvignac, C. Pichot, P. Ratajczak, and P. Brachat. A new non-planar Vivaldi antenna. In *2005 IEEE Antennas and Propagation Society International Symposium*, volume 1A, pages 565–568, 2005.

[11] F. W. Yao, S. S. Zhong, and X. X. L. Liang. Experimental study of ultra-broadband patch antenna using a wedge-shaped air substrate. *Microwave and Optical Technology Letters*, 48(2):218–220, 2006.

[12] R. Ericsson. WP 2.3-1 wideband antenna radiators-TEM horn. Technical report, Swedish Defense Research Agency, 2004.

[13] K. H. Chung, S. H. Pyun, S. Y. Chung, and J. H. Choi. Design of a wideband TEM horn antenna. In *2003 IEEE International Symposium and Meeting on Antennas and Propagation and USNC/URSI National Radio Science*, pages 229–232, 2003.

[14] R. T. Lee and G. S. Smith. A design study for the basic TEM horn antenna. *IEEE Antennas and Propagation Magazine*, 46(1):86–92, 2004.

[15] R. T. Lee and G. S. Smith. A design study for the basic TEM horn antenna. In *2003 IEEE International Antennas and Porpagation Society International Symposium*, volume 1, pages 225–228, 2003.

[16] D. A. Kolokotronis, Y. Huang, and J. T. Zhang. Design of TEM horn antennas for impulse radar. In *1999 High Frequency Postgraduate Student Colloquium*, pages 120–126, 1999.

[17] S. Licul. *Ultra-wideband antenna characterization and measurement.* PhD thesis, Virginia Polytechnic Institute, USA, 2004.

[18] S. Herrero Arias and J. E. Fernández del Río. Optimización de la directividad de antenas quasi-Yagi sobre FR4 para aplicaciones WiFi. In *XX Simposio Nacional de la URSI (in Spanish)*, 2005.

[19] J. Huang and A. C. Densmore. Microstrip Yagi array antenna for mobile satellite vehicle application. *IEEE Transactions on Antennas and Propagation*, 39(7):1024–1030, 1991.

[20] W. R. Deal, N. Kaneda, J. Sor, Y. Qian, and T. Itoh. A new quasi-Yagi antenna for planar active antenna arrays. *IEEE Transactions on Microwave Theory and Techniques*, 48(6):910–918, 2000.

[21] L. C. Kretly and A. S. Ribeiro. A novel tilted dipole quasi-Yagi antenna designed for 3G and Bluetooth applications. In *Proceedings of the 2003 SBMO/IEEE MTT-S International Microwave and Optoelectronics Conference*, volume 1, pages 303–306, 2003.

[22] H. J. Song, M. E. Bialkowski, and P. Kabacik. Parameter study of a broadband uniplanar quasi-Yagi antenna. In *3th International Conference on Microwave, Radar and Wireless Communications*, volume 1, pages 166–169, 2000.

[23] N. Kaneda, Y. Qian, and T. Itoh. A broad-band microstrip-to-waveguide transition using quasi-Yagi antenna. *IEEE Transactions on Microwave Theory and Techniques*, 47(12):2562–2567, 1999.

[24] S. Y. Chen and P. Hsu. Broadband microstrip-fed modified quasi-Yagi antenna. In *IEEE/ACES International Conference on Wireless Communications and Applied Computational Electromagnetics*, pages 208–211, 2005.

[25] H. K. Kan, A. M. Abbosh, R. B. Waterhouse, and M. E. Bialkowski. Compact broadband coplanar waveguide-fed curved quasi-Yagi antenna. *IET Microwave Antenna Propagation*, 1(3):572–574, 2007.

[26] Y. Qian, W. R. Deal, N. Kaneda, and T. Itoh. A uniplanar quasi-Yagi antenna with bandwidth and low mutual coupling characteristics. In *IEEE Antennas and Propagation Society International Symposium*, pages 924–927, 1999.

[27] C. A. Balanis. *Antenna Theory: Analysis and Design*. John Wiley & Sons, 3rd edition, 2005.

[28] M. A. Peyrot-Solis. *Investigación y Desarrollo de Antenas de Banda Ultra Ancha (in Spanish)*. PhD thesis, Center for Research and Advanced Studies of IPN, Department of Electrical Engineering, Communications Section, Mexico, 2009.

[29] M. A. Peyrot-Solis, G. M. Galvan-Tejada, and H. Jardón-Aguilar. Proposal and development of two directional UWB monopole antennas. *Progress in Electromagnetics Research C*, 21:129–141, 2011.

[30] G. H. Brown and O. M. Woodward Jr. Exexperimental determined radiation characteristics of conical and triangular antennas. *RCA Review*, 13(4):425–452, 1952.

[31] M. A. Peyrot-Solis, G. M. Galvan-Tejada, and H. Jardón-Aguilar. A novel planar UWB monopole antenna formed on a printed circuit board. *Microwave and Optical Technology Letters*, 48(5):933–935, 2006.

8

Current Tendencies and Some Unresolved Problems

CONTENTS

8.1 UWB Antennas Today

Over a decade has passed since the United States Federal Communication Commission specified the frequency band for ultra wideband applications; since then this technology has become an area of study not only in the research community, but also for academic and industry sectors. In addition, like many other technological developments which were developed in military environments, UWB designs are now also available for civilian implementations. The importance of the subject is illustrated by the vast quantity and variety of books, journal papers, and conference presentations on the topic that can be found in the open literature. In the area of antennas, for instance, we can cite a number of books directly dedicated to the fundamentals and design of UWB radiators (e.g., [1–3]), as well as other texts that indirectly address aspects related to the UWB antenna principles [4].

In studying this material, it is apparent that there are different proposals for UWB antenna designs, from the early volumetric structures to the recent planar and planarized profile radiators. Some of these are reported to have omnidirectional radiation characteristics, so they may be implemented in the emerging UWB communications, with their tendency toward high mobility and portability. Directive designs have also attracted the attention of the research community for military applications, spectrum monitoring, imaging, and radar, although they have received less attention in comparison to their omnidirectional counterpart.

In any case, one of the main parameters is the antenna bandwidth. Indeed, this term forms the basis of different concepts in the theory and design of antennas. However, in an ultra wideband context, it is difficult to find a single definition of bandwidth. In general, the bandwidth is usually referred to as the impedance matching response that a particular antenna presents. Nevertheless, the phase character and the radiation pattern behavior cannot be neglected when we are addressing non-resonant devices such as the UWB antennas.

From what has been discussed, one can highlight the following aspects as important to be covered, mainly for planarized and planar UWB antennas: In the case of the former, it is preferable for antennas to have a simple construction, be low cost, and to have relatively small dimensions that have a reflection coefficient magnitude less than or equal to -10 dB in the whole operating range, with a gain greater than 2.5 dBi in the direction of their main lobule, and that their operating band have wider lower and upper limits in 1 GHz of the frequency bands allocated by the FCC for short-range UWB communications. For UWB planar monopole antennas, it is necessary to have both omnidirectional and directional designs, of relatively small size, that do not require of the use of a balun, and that have a relatively stable radiation pattern. In cases where a particular application requires UWB bandwidths, but has power levels higher than those supported by the current UWB planarized antennas, new designs are required.

8.2   Impedance Matching, Phase Linearity and Radiation Pattern

In following the book's guidelines, it has been stressed that the design of UWB antennas involves three fundamentals of analysis. First of all, the antenna must be matched, which is evaluated through the magnitude of the reflection coefficient. As already pointed out, a value of $|\Gamma| < -10$ dB is usually accepted as the performance threshold in such a way that the curve of $|\Gamma|$ must be below this limit for the desired operational frequency to be reached. Care must be taken when evaluating claims of some authors that their antennas can operate over a certain bandwidth, which is related to a $VSWR = 2.5$ (or even higher) instead of $VSWR = 2.0$, for instance, because that implies larger return losses in the antenna.

The second performance parameter is the phase response. As explained in Chapter 5, essentially there are two approaches followed by different authors to evaluate this. One approach is that based on the antenna reflection coefficient, and the other is that which considers the antenna transfer function. It is worth noting that several authors report the phase behavior of certain antennas by installing a UWB system, where depending on the received pulse characteristics, the transfer function is derived.

The last parameter corresponds to the variation of the radiation pattern as a function of frequency. The aim is to conserve the pattern shape, gain, front to back lobe ratio (for directive radiators), etc., through the operational bandwidth. This condition is a hard constraint because, since UWB antennas are non-resonant devices, they operate over a wide frequency span, which introduces changes in the currents distribution as a function of the frequency. So, this situation produces variations in the radiation pattern, and hence it is suggested that a transverse wave response of the antenna be sought. Now, another definition of the bandwidth is related to this parameter, which is the frequency interval where the pattern conserves its characteristics. Thus, a given antenna could be well-matched in a particular bandwidth (for $|\Gamma| <$ -10 dB), having an acceptable phase response, but its radiation pattern may hold in a narrower bandwidth.

Therefore, one of the most difficult challenges in designing UWB antennas is to fulfill simultaneously these three performance requirements. Occasionally, one of them could be relaxed (for example, if the antenna is not intended to transmit data at a high rate, as in spectrum monitoring, the phase response could tolerate some non-linearities), but the baseline must always be the condition of a well-matched antenna.

8.3   Directional UWB Antennas

As has been seen, there is little attention given to directional UWB antennas in the open literature. Antennas with this characteristic are important to study, provided that they are not only interesting for military applications, but also for radar, imaging, and telecommunications for civilian applications.

One example of the application of directional UWB antennas is the emerging area of body area networks (introduced in Section 8.6 below), where evaluations have been carried out to compare the performance of distinct antennas (both omnidirectional and directional) in order to assess their inclusion in these networks. There is much pending work in this area of research.

Another niche of UWB directional antennas can be found in the context of spectrum monitoring. One of the tasks associated with this activity is known as *direction finding, DF*, where the radiation pattern of an antenna is focused in a particular direction, to detect possible interference sources. Naturally, the radiation characteristics of antennas used in this application correspond to directional patterns. Now, the whole system is usually made up of more than one antenna, in such a way that each is in charge of a specific interval of frequencies. Thus, if a single antenna is able to cover a wider frequency span by exploiting the UWB capabilities, it is also possible to implement a DF-UWB antenna. Therefore, ultra wideband radiators specifically designed for this objective are needed. One possible design in this area is the TEM horn antenna introduced in Chapter 3, and evaluated in Chapter 7. Of course, improvements by means of the scaling principle described in Chapter 6, for instance, could be achieved if looking to apply the horn antenna to spectrum monitoring.

8.4   UWB Antenna Arrays

As explained in Section 2.2.5, structures based on antenna arrays represent one of several possibilities to attain a directional radiation pattern. However, some examples of antennas based on these structures, like the popular log-periodic antenna, are highly dispersive (see Section 5.5.1), which is a limitation for UWB applications. One attempt at a UWB array is the quasi-Yagi antenna studied in Chapter 7. Nevertheless, this radiator presents the narrowest bandwidth within which the phase behaves with a non-linear character. The advantage of this design is its low profile, which makes it attractive for aircrafts.

Another application of antenna arrays is for implementing a diversity mechanism to mitigate the effects of multipath fading caused by diverse scattering of objects both outdoors and indoors. If a UWB communication system is being planned using an antenna array, it is necessary to carefully study the interactions of the array elements involved in producing a certain response, and how their structures could modify it.

In [5], for instance, a study on UWB antenna arrays is carried out. Specifically, the classical narrowband ULA (see Chapter 2) is explored to become useful in UWB applications. Usually the wavelength is determined from a central frequency in such a way that the inter-spacing of the array elements is calculated as $\lambda/2$. However, as Gentner et al. mentioned (and we have pointed out in Chapter 4), it is not possible to define a central frequency in UWB systems. Thus, authors take as a reference the lowest frequency of the UWB band (in this particular case from 3 to 6 GHz) and find that the performance of the ULA is degraded through the band. Thus, a seven element non-uniform linear array (NULA) is proposed following a geometric progression for the inter-element spacing. Results are favorable not only in antenna array performance, but also a size reduction around 23% is achieved in comparison with the classical ULA. This feature is of particular interest for applications where dimensions of the antenna become critical.

In the past two decades, a branch of the development of antenna arrays evolved into what is known as *Multiple Input-Multiple Output* (MIMO) technology which has matured and has been applied to several communication systems. In [6] the authors study and compare antenna arrays for MIMO applications in the UWB bandwidth like high-data rate wireless communications, high precision localization, and radar imaging, among others. In essence, a ULA is compared with an array proposed by the authors called an *interleaved uniform rectangular array* (iURA). The objective of the iURA is to reduce mutual coupling between elements in order to conserve the omnidirectional shape of the radiation pattern through a UWB bandwidth such that the effects due to dense multipath produced in an indoor environment can be mitigated. The antenna proposed in [6] operates between 2.2 and 7.5 GHz with a gain of approximately 3 dBi.

8.5   Interference

Without doubt an important aspect of the operation of UWB systems is their electromagnetic compatibility due to the very broad range of frequencies over which signals are transmitted, and therefore all the bands that they share with other wireless networks (e.g., WLANs). There are different approaches to mitigate interference effects between UWB systems and other networks. First of all, remember that UWB systems work on an energy ultra short pulse basis and hence UWB pulses are far harder to intercept. In other words, the operation characteristics of the UWB systems made themselves robust against interference. Nevertheless, their electromagnetic compatibility is not infinite. Then, in order to provide these systems with an extra protection, some considerations can be taken on the design of their antennas, such as achieving a notched response on the magnitude of the reflection coefficient. This approach has been studied by several authors for many years (see for example, the CPW-fed planar wideband antenna and the microstrip slot antenna with fractal tuning stub discussed in Chapter 3). Designs of UWB antennas oriented to interference avoidance are so current that many research works can be found in the open literature. For example, recently different UWB antenna structures based on a dielectric resonator antenna (DRA) are explored in [7]. One of the applications is exactly for interference rejection in WLANs due to the overlapped bands of these wireless networks with the UWB band assigned by the FCC. In this case two UWB planarized DRA were designed; on the one hand, a single strip structure that produces a notch band at 5.15–5.825 GHz and, on the other hand, a two separate strips antenna, which introduces notch bands at 5.15–5.35 GHz and 5.725–5.825 GHz.

8.6   Body Area Networks

Body area networks (BANs) are wireless networks used, mainly, to monitor some bodily functions of human beings. Basically a set of sensors are attached to specific parts of the body (e.g., chest for electrocardiogram, arm for glucose levels, wrist for blood pressure, ankles, knees and arms for movement, etc.) in such a way that a person can be monitored at a distance, either for a routine medical check-up, or in an emergency situation in which a patient requires immediate attention [8]. Even a simple fall could be monitored through a system of movement pattern recognition for people who live alone. For non-medical scenarios, BANs have also attracted the attention of the entertainment and gaming sector (computer games, dance lessons, etc.), as well as for defense [8]. In any case, these sensors require collected information to be sent to a hub, which is in charge of forwarding the information to an access point, through which an Internet connection allows the data to be carried to the final destination. Figure 8.1 shows a representation of a wireless BAN (for a health care case), where the Internet connections are marked as well.

Three types of links make up the wireless channel of a BAN [9]: On-body links, off-body links and in-body links. Basically they are related to the position of the transmitter and receiver devices. For on-body links both the transmitter and the receiver are mounted on the same body, whereas for off-body links, both terminals are not on the same body. In the case of in-body links, only one of the terminals is embedded in the body.

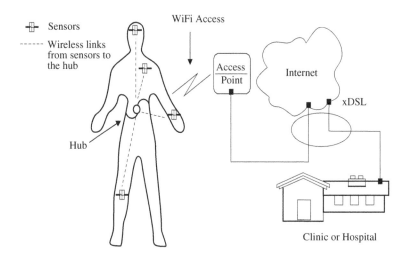

FIGURE 8.1
Representation of a BAN.

TABLE 8.1

Frequency bands considered for wireless body area networks

Frequency (MHz)	Acronym
402 − 405	MedRadio
433.05 − 434.79	General telemetry
608 − 614, 1395 − 1400, 1427 − 1432	WMTS
868 − 870	General telemetry
902 − 928	ISM
2400 − 2483.5	ISM
5725 − 5850	ISM
4200 − 4800, 7250 − 8500	UWB

WMTS: wireless medical telemetry service; ISM: industrial, scientific and medical

On the other hand, different frequency bands have been considered for the operation of wireless BANs, as shown in Table 8.1 [8]. The fundamental reason for using microwave frequencies lies in the fact that BANs require relatively small antennas to be attached to the body or in fabrics.

Since some bands for BANs correspond partially to the FCC UWB spectrum, and provided that UWB antennas are designed to transmit relatively low power levels, they have been considered as attractive candidates for wireless BANs. Thus, recent works have analyzed diverse UWB antennas used in BANs, from the classical ones formed on printed circuit boards [10–15], textile antennas based on copper, nickel, and silver nylon fabric [16], wearable low profile antenna with orthogonal polarization [9], up to the recent double ring finger antenna [17]. It is worth noting that the proposals of [9–15] are designed to operate in the interval of 3−6 GHz, whereas the double ring antenna is considered for the upper UWB band (7.25−10.25 GHz), and the textile radiators are proposed for the whole FCC UWB band.

Some works also evaluate the effect of the body on the UWB antennas, because each tissue presents certain electrical properties [12, 14, 15]. Three general types of tissues can be identified for these evaluations: skin, fat, and muscle (with their corresponding dielectric constant ε_{r_s}, ε_{r_f} and ε_{r_m}, respectively), which are simply simulated in a layered or stack model as shown in Figure 8.2.

For example, a planarized omnidirectional antenna is designed and simulated in [18] taking into the consideration the electrical properties of three different tissues of a human head: skin, fat and bone.[1] Basically, the idea is that the proposed antenna is implanted under skin, but over fat and bone. Simulations were conducted for this stack scheme and for different values of the electric constants depending on the frequency. Results show that this antenna presents a good impedance matching between 3–10 GHz, which provides

---

[1]Please note that Yazdandoost [18] does not take exactly all layers shown in Figure 8.2, where muscle is included, but the stack approach is followed where bone is introduced.

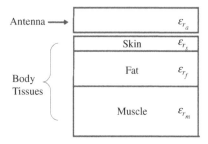

FIGURE 8.2
Layered representation of main tissues considered for simulations.

better capabilities of data transfer than the current band of MICS (Medical Implant Communication Service) from 402–405 MHz. Practically an unchanged radiation pattern as a function of the frequency is also reported.

Then, depending on the electrical characteristics of the body and the tissue thickness, the antenna performance could be affected by the human body. Therefore, antenna designs in these lines of research should take into account this fundamental aspect. The impact of the distance between an antenna and a human body is analyzed in [19] for WBAN applications with on-body antennas. Two UWB planarized antennas, loop and dipole, are considered for simulations and measurements varying the aforementioned distance in an interval between 1-12 GHz. The comparison parameter is the magnitude of the reflection coefficient, where a free space environment is taken as a reference. According to the results reported by the authors, neither antenna covers the whole band allocated for UWB systems since the magnitude of the reflection coefficient presents strong variations mainly to shorter distances. However, it seems that 20 mm can be considered an acceptable distance where the antennas performance can be relatively stable. An interesting analysis is also included in terms of the radius of the Wheelers radiansphere, from where it is explained the behavior of antennas very near of the body and how they can present characteristics of electrically small antennas.

8.7   Radar: Medical Imaging and Others

Radar technology has been developed for many years in a wide variety of applications. The arrival of UWB antenna designs also has reached radar systems and different research lines on the subject have been opened around the world. Due to the importance and current problems of the topic, the first application addressed in this section is medical imaging. Later on other examples of the UWB radar are described.

High resolution techniques like microwave imaging provide an attractive solution to several medical applications where it is necessary to detect tumors and therefore at microwave and higher frequencies the tissue penetration is feasible. In addition, by comparison to traditional X-ray techniques, microwave imaging treatment is less intrusive on patients, who are also exposed to lower levels of radiation. In contrast, X-ray based treatment implies high doses of ionizing radiation, which impacts on a much more limited usage. Another point is that, in the particular case of breast cancer detection, at X-ray frequencies, poorer contrast results between healthy and diseased tissues when patients are younger [20]. The penetrating capability of microwave imaging avoids painful breast compression.

Since perhaps the most representative case is for breast cancer, let us explain microwave imaging and the use of UWB antennas in the context of this disease. In this matter, there exist two main techniques related to breast cancer detection [20]: *microwave tomography* and *radar-based imaging*. In the former, an image of the distribution of the dielectric constant in the breast is reconstructed through an inverse scattering problem. In the latter the aim is to identify the presence and location of scattering breast tumors using a radar basis. In essence, the radar-based imaging is shown in Figure 8.3, where a very simple scheme depicts the process: First, a microwave signal is transmitted to cover the area under analysis. If a breast tumor is in the area, this target reflects part of energy in such a way that the receive antenna intercepts it. Finally, the information is passed to a processing unit where a reconstruction of the whole scene and the location of the tumor is identified.[2]

Naturally more than one measurement has to be carried out at different positions in the area of interest in order to have a good enough spatial resolution. The above is related to what is known as *space-time beamforming* techniques, which find their fundamentals in the joint use of antenna arrays and digital signal processors (for this reason this technique is also known as *digital beamforming*). This technology has become so mature that its popularity has reached diverse applications, including radar imaging for breast cancer detection. The main advantage of digital beamforming techniques is that more than one target can be identified at once depending on the antenna array capabilities like structure, type of elements, inter-element spacing, etc., and the source separation algorithm implemented on the DSP. Details on this theme are outside the scope of this book, but the interested reader can consult [21] and references within for a review of works studied by many authors from diverse points of view.

Regarding the UWB antennas used in imaging for breast cancer detection (or some other type of medical detection), it is worth mentioning that one

---

[2]Please note that separate antennas and transmitter and receiver stages are illustrated in Figure 8.3, which is not necessarily the practical case. It was done so here to simplify the explanation.

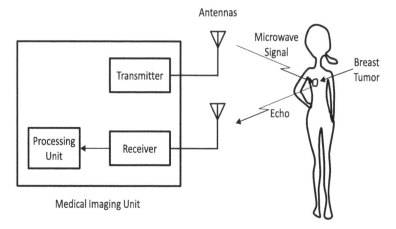

FIGURE 8.3
Microwave radar-based imaging scheme.

of the topics to note in simulations is how to model the breast (or another specific part of the human body) in terms of the dielectric properties and then to introduce a proper phantom (see for example [20, 22, 23]). In [24], for instance, a directional antenna array operating approximately from 1 to 8 GHz for pulse-based systems suitable for detection of breast cancer and urine in human bladder is proposed. This antenna was simulated using phantom materials with different dielectric constants in order to emulate human tissues. On the other hand, Lai et al. [22] study and present a wide set of experimental results with heterogeneous breast phantoms. Their analysis is carried out in the time domain using a Gaussian pulse generator (transmitter) and a real-time oscilloscope (receiver). The UWB antennas used for the testbed are 3 cm width and 4 cm height with a gain of 11 dBi for a impedance bandwidth from 2.4–12 GHz. According to their observations, care should be taken on the hardware for detecting small tumors (millimeters size) when a higher dielectric permittivity is considered.

Now, the main parameter to evaluate in radar imaging based on UWB antennas is the temporal response of the pulses. In order to have a high resolution response and therefore a reliable diagnostic, low distortion of the pulses should be attained. The antenna array proposed in [24], for instance, presents this behavior and hence it can be used for medical diagnostics. Another example of is the H-shaped UWB planarized DRA designed by [7] for breast cancer detection. The authors claim that the advantage of the DRA used for this type of application is the ability to use it without a matching medium. This is so because the dielectric constant of the DRA is very close to the permittivity of the fatty tissue. The main features of this antenna are: constant gain, high efficiency, a bandwidth very close to that assigned by the FCC for UWB, and a group delay of the order of 0.15 ns for the operating frequency.

Let us now describe a couple of different applications of UWB radar and their antennas designed. First an interesting use of radar technique is for what is known as *Ground Penetrating Radar* (GPR), where detection of small and shallow objects buried in the ground, hidden tunnels, cables, pipes, and landmines is the central objective. Then, very accurate time and broadband frequency domain responses are needed. In addition, a directional radiation pattern that is relatively constant as a function of frequency is also required. Thus, a directional planarized CPW fed antenna used for GPR designed between 0.4 and 3 GHz is proposed in [25]. Although this antenna does not cover the FCC UWB band, it does fulfill the condition of a bandwidth larger than 500 MHz for which a gain between 5 and 8.5 dBi is approximately achieved.

The final application to be addressed in this section is the classical radar usage for aircraft. In this subject, a migration of a combined structure made up by a horn TEM antenna and an inductive loop antenna is studied in [26] as a low profile antenna suitable for radar installed on an airborne surface. As a result of this migration, the authors propose a directional antenna called electrically narrow very low profile (ENVELOPE). The main application of this antenna is for radar, which implies that a directional radiation pattern is required, a feature achieved by the ENVELOPE structure. In order to improve the antenna characteristics, the ENVELOPE antenna was passed through a tuning process. A couple variants of the ENVELOPE structure based on a circular array were explored for diverse applications, both with a polarization and a radiation pattern with satisfactory results. Although all these antennas are not in the band assigned by the FCC for UWB applications, but are from 1–3 GHz approximately, they present a bandwidth of more than one octave, which belongs to one of the definitions of UWB discussed in the early part of this book.

8.8   USB Dongle and Access Point

The extension of Universal Serial Bus (USB) technology to a wireless mode has attracted much the attention in recent years. Among the wide variety of applications of USB technology is that of a connectivity tool in data exchange between various electronic devices making up a Personal Area Network (e.g., diverse consumer electronic devices or personal computers, laptops at home used to share video, photos, etc.). Nevertheless, a main restriction is to achieve low profile antennas for USB dongle dimensions that also operate in a very wide bandwidth in such a way that a huge data stream can be downloaded. Thus, the wireless USB approach implemented over a UWB technology could provide short range and high data communications. Two examples of design can be cited here where low profile antennas are explored for this application. First, an omnidirectional monopole antenna operating from 2.8 to 13.3 GHz for a USB dongle application was proposed and designed in [27]. Second, a compact U-shaped planarized UWB antenna is designed and studied in [28], which presents quasi-omnidirectional radiation characteristics in a simulated impedance bandwidth between 3–8 GHz (although the authors claim that a 3.1–10.6 GHz range is covered in free space). Both present a compact size suitable for USB dimensions.

In contrast, a relatively larger antenna was designed by [7] which could be applied as an access point for UWB communications. This three dimensional antenna is made up of a wrapped structure above a ground plane. In the context of this application, it presents an omnidirectional pattern with a quasi-constant gain through the operational bandwidth (from 4 to 14 GHz approximately). Thus, a classical omnidirectional coverage could be suitable to conform to a high data rate wireless network.

8.9   Computational Aspects

The need of multiple-precision arithmetic to solve computational electromagnetics problems is addressed in [29], where five categories of problems are identified:

1. Generation of special mathematical functions applicable in electromagnetic theory.

2. Solving ill-conditioned linear systems of equations in electromagnetic computations.

3. Simulation of large-scale radio-frequency circuits.

4. Difficult full-wave simulations of electromagnetic phenomena.

5. Experimental electromagnetic theory

Another problem is without doubt the stability of the numerical methods, which is always a matter of interest not only for electromagnetic applications, but for all areas of knowledge. In particular, Aksoy and Ösakin [30] present a stability analysis of the known Finite Difference Time Domain method (to be addressed in Chapter 9), provide its relatively direct implementation and good approximation.

In the particular case of UWB antennas for WBANs, simulations should include body phantoms in order to consider the effects of individual tissues on the antenna performance. However, as Cara et al. stated [9], the relationship of the relative proportion between antenna and phantom dimensions can have impacts on the computational resources.

8.10 Wider and Wider Bandwidths

Finally, it is important to note the need to have antennas whose bandwidths are wider than those specified by the FCC for UWB applications. The reason for this is that a single antenna can cover a wide range of frequencies for spectrum monitoring purposes (UWB directional antennas were considered in Section 8.3 as a possibility for direction finding, but also UWB omnidirectional radiators could be useful for other electromagnetic spectrum alertness tasks). A possible approach to providing this capability is by using some guidelines as given in Chapter 6, whereby the scaling factor was introduced in order to design the antenna in a frequency band different from that of its original structure.

Naturally, an electromagnetic compatibility analysis must be conducted when a new design is proposed, mindful of the FCC spectral masks or, in the case of bandwidths wider than that of regulatory organisms, aware of possible interference caused to systems or equipment operating in a common band.

Bibliography

[1] H. Schantz. *The Art and Science of Ultra Wideband Antennas.* Artech House, Norwood, MA, 2005.

[2] B. Allen, M. Dohler, E. E. Okon, W. Q. Malik A. K. Brown, and D. J. Eduards, editors. *Ultra-Wideband Antennas and Propagation for Communications, Radar and Imaging.* John Wiley & Sons, West Sussex, UK, 2007.

[3] D. Valderas, J. I. Sancho, D. Puente, C. Ling, and X. Chen. *Ultrawideband Antennas, Design and Applications.* Imperial College Press, London, UK, 2011.

[4] Z. N. Chen and M. Y. W. Chia. *Broadband Planar Antennas: Design and Applications.* Jonh Wiley & Sons, Sussex, England, 2006.

[5] P. K. Gentner, G. S. Hilton, M. A. Beach, and C. F. Mecklenbräuker. Characterisation of ultra-wideband antenna arrays with spacings following a geometric progression. *IET Communications*, 6(10):1179–1186, 2012.

[6] X.-S. Yang, J. Salmi, A. F. Molisch, S.-G. Qiu, and S. Sangodoyin. Trapezoidal monopole antenna and array for UWB-MIMO applications. In *2012 International Conference on Microwave and Millimeter Wave Technology*, volume 1, pages 1–4, 2012.

[7] A. A. Kishk, X. H. Wu., and S. Ryu. UWB antenna for wireless communication and detection application. In *2012 IEEE International Conference on Ultra-Wideband*, pages 72–76, 2012.

[8] M. Patel and J. Wang. Applications, challenges, and prospective in emerging body area networking technologies. *IEEE Wireless Communications*, 17(1):80–88, 2010.

[9] D. D. Cara, J. Trajkoviki, R. Torres-Sánchez, J.-F. Zürcher, and A. K. Skrivervik. A low profile UWB antenna for wearable applications: the tripoid kettle antenna (TKA). In *2013 7th European Conference on Antennas and Propagation*, pages 3257–3260, 2013.

[10] M. Klemm, I. Z. Kovacs, G. F. Pedersen, and G. Tröster. Comparison of directional and omni-directional UWB antennas for wireless body area network applications. In *18th International Conference on Applied Electromagnetics and Communications*, pages 1–4, 2005.

[11] A. Alomainy, Y. Hao, C. G. Parini, and P. S. Hall. Comparison between two different antennas for UWB on-body propagation measurements. *IEEE Antennas and Wireless Propagation Letters*, 4:31–34, 2005.

[12] T. S. P. See and Z. N. Chen. Experimental characterization of UWB antennas for on-body communications. *IEEE Transactions on Antennas and Propagation*, 57(4):866–874, 2009.

[13] G. Alpanis, C. Fumeaux, J. Frönlich, and R. Vahldieck. A truncated conical dielectric resonator antenna for body-area network applications. *IEEE Antennas and Wireless Letters*, 8:279–282, 2009.

[14] K. Y. Yazdandoost and K. Hamaguchi. Very small UWB antenna for WBAN applications. In *5th International Symposium on Medical Information & Communication Technology*, pages 70–73, 2011.

[15] L. Lizzi, G. Oliveri, F. Viani, and A. Massa. Synthesis and analysis of a monopole radiator for UWB body area networks. In *2011 IEEE-APS Topical Conference on Antennas and Propagation in Wireless Communications*, pages 78–81, 2011.

[16] M. Klemm and G. Troester. Textile UWB antennas for wireless body area networks. *IEEE Transactions on Antennas and Propagation*, 54(11):3192–3197, 2006.

[17] H. Goto and H. Iwasaki. A low profile monopole antenna with double finfer ring for BAN and PAN. In *2011 International Workshop on Antenna Technology*, pages 227–230, 2011.

[18] K. Y. Yazdandoost. UBW antenna for body implanted applications. In *Proceedings of the 42th European Microwave Conference*, pages 932–935, 2012.

[19] T Tuovinen, T. Kumpuniemi, K. Y. Yazdandoost, M. Hämäläinen, and J. Iinatti. Effect of the antenna-human body distance on the antenna matching in UWB WBAN applications. In *2013 7th International Symposium on Medical Information and Communication Technology*, pages 193–197, 2013.

[20] I. Ünal, B. Türetken, K. Sürmeli, and C. Canbay. An experimental microwave imaging system for breast tumor detection on layered phantom model. In *2011 XXXth URSI General Assembly and Scientific Symposium*, pages 1–4, 2011.

[21] S. Haykin and K. J. Ray Liu, editors. *Handbook on Array Processing and Sensor Networks*. Wiley-IEEE Press, 2009.

[22] J. C. Y. Lai, C. B. Soh, E. Gunawan, and K. S. Low. UWB microwave imaging for breast cancer detection – experiments with heterogeneous breast phantoms. *Progress in Electromagnetics Resarch M*, 16:19–29, 2011.

[23] M. A. Shahira Banu, S. Vanaja, and S. Poonguzhali. UWB microwave detection of breast cancer using SAR. In *2013 International Conference on Energy Efficient Technologies for Sustainability*, pages 113–118, 2013.

[24] X. Li, J. Yan, M. Jalilvand, and T. Zwick. A compact double-elliptical slot-antenna for medical applications. In *6th European Conference on Antennas and Propagation*, pages 36–3680, 2011.

[25] P. Cao, Y. Huang, and J. Zhang. A UWB monopole antenna for GPR application. In *6th European Conference on Antennas and Propagation*, pages 2837–2840, 2011.

[26] A. Elsherbini and K. Sarabandi. ENVELOPE antenna: a class of very low profile UWB directive antennas for radar and communication diversity applications. *IEEE Transactions on Antennas and Propagation*, 61(3):1055–1062, 2013.

[27] C.-M. Wu, Y.-L. Chen, and W.-C. Liu. A compact ultrawideband slot patch antenna for wireless USB dongle application. *IEEE Antennas and Wireless Propagation Letters*, 11:596–599, 2012.

[28] E. K. I. Hamad and A. H. Radwan. Compact UWB antenna for wireless personal area networks. In *2013 Saudi International Electronics, Communications and Photonics Conference*, pages 1–4, 2013.

[29] T. P. Stefański. Electromagnetic problems requiring high-precision computations. *IEEE Antennas and Propagation Magazine*, 55(2):344–353, 2013.

[30] S. Aksoy and M. B. Özakin. A new look at the stability analysis of the finite-difference time-domain method. *IEEE Antennas and Propagation Magazine*, 56(1):293–299, 2014.

9

<hr style="height:4px; background:black; border:none;" />

Numerical Methods for Electromagnetics

CONTENTS

In Chapter 2 some fundamentals of antennas were introduced in order to understand and address the problem of UWB antennas. Due to the complexity of solving analytically the electromagnetic equations related to antennas (either narrowband, wideband or ultra wideband), the present chapter is dedicated to the subject of solving them numerically, through which the reader is provided with some ideas of the background of the simulation packages used to this aim. As will be seen in Section 9.1, the essence of the electromagnetic theory formulated by James Clerk Maxwell can be outlined by a set of partial differential equations. There exist many references (including symposiums) addressing solution methods for this type of equation (see [1–5] for instance). Although it is not, of course, the aim of this chapter to present a treatise on differential equations, but to point out some particular aspects of them related to the Maxwell's equations, Appendix B is just given as a quick reference. Finally, it is important also to note that the complexity of the topic of numerical methods usually requires a wide treatise like in [6–11] to mention just a few. In spite of the fact that only a brief exposition and the background for these methods are addressed in this chapter, they are included as a part of this book due to the relevance of the theme.

9.1   Maxwell's Equations

9.1.1   Basic field laws

As is well known, an antenna can be seen as a medium through which the guided energy traveling in a transmission line or waveguide is transformed into unguided electromagnetic (EM) energy to be propagated to space. So, one can consider any antenna as a source of an EM field. On the other hand, the source of an EM field is a density of current variant with time ($\mathbf{J}$), which is associated to a charge density also variant with time (ρ). Both quantities are related to each other through the Continuity Equation or Conservation of Charge Law [12]

$$\nabla \cdot \mathbf{J} = -\frac{\partial \rho}{\partial t} \tag{9.1}$$

with ∇ as the differential operator (see Appendix A). Thus, the aim of the antenna analysis is to determine the expressions of electric and magnetic fields as a function of charge and current distribution existing in it. From these expressions it is then possible to derive some important antenna parameters such as the radiated power density. The analysis of EM fields (and therefore of the electric and magnetic fields) is based on Maxwell's equations as given by:

$$\nabla \times \mathbf{E} + \mu_0 \frac{\partial \mathbf{H}}{\partial t} = 0 \tag{9.2}$$

$$\nabla \times \mathbf{H} - \varepsilon_0 \frac{\partial \mathbf{E}}{\partial t} = \mathbf{J} \tag{9.3}$$

$$\nabla \cdot \varepsilon_0 \mathbf{E} = \rho \tag{9.4}$$

$$\nabla \cdot \mu_0 \mathbf{H} = 0 \tag{9.5}$$

where $\mathbf{E}$ represents the electrical field, $\mathbf{H}$ the magnetic field, ε_0 is the permittivity of vacuum, and μ_0 is the permeability of vacuum. Please note that these four equations plus Equation (9.1) form the complete set of Maxwell's equations.[1] The derivation from the integral form to Equations (9.1)–(9.5) can be found in several textbooks like [12]. In attribution to persons who gave physical origin to the Maxwell's equations, Equations (9.4) and (9.5) are usually referred to as Gauss' Law for electric and magnetic fields, respectively, whereas Equation (9.2) corresponds to the Faraday's Law, and Equation (9.3) is Ampere's Law. It is worth noting that the products $\varepsilon_0 \mathbf{E}$ and $\mu_0 \mathbf{H}$ in the Gauss' Laws are related with the flux densities of electric and magnetic fields, respectively, through

$$\mathbf{D} = \varepsilon_r \varepsilon_0 \mathbf{E} \tag{9.6}$$

$$\mathbf{B} = \mu_r \mu_0 \mathbf{H} \tag{9.7}$$

where $\varepsilon_r = \varepsilon/\varepsilon_0$ and $\mu_r = \mu/\mu_0$, with ε the permittivity of the medium and μ the permeability of the medium and therefore if the medium is vacuum, $\varepsilon_r = 1$ and $\mu_r = 1$.

It is important to remark that both Equations (9.2) and (9.3) present a relationship between $\mathbf{E}$ and $\mathbf{H}$ fields, which is one of the fundamentals of the electromagnetic theory. Mathematically, Equations (9.2)–(9.5) represent a set of partial differential equations in the three spatial dimensions plus the time dimension. Usually, the solution of them can analytically be determined for a few "single" cases only. It is due to the fact that an EM source is not isolated

---

[1]Actually Maxwell presented a very wide set of twenty equations in his original work, and years later he did an extraction to these famous four equations. Many persons were so interested in the modeling of EM fields based on the Maxwell's treatise that different mathematical approaches were developed during the late nineteenth century, until the arrival of the present vector notation attributed to Heaviside [13, 14].

in free space and therefore the interaction with other materials (conductors or dielectrics) should be taken into account. In other words, the boundary between materials and free space has to be included in the development of a solution of Maxwell's equations, hence we have at hand a set of partial differential equations with a boundary value problem to be solved.

Depending on the problem, the equations' system could be non-homogeneous, for which a general solution is needed. Naturally one is looking for a unique solution to the set of Maxwell's equations, then we have to be supported on something for that purpose. The above is given by the *Uniqueness Theorem*, which states [8]:

Uniqueness Theorem

"In a region V completely occupied with dissipative media, a harmonic field $(\mathbf{E}, \mathbf{H})$ is uniquely determined by the impressed currents in that region plus the tangential components of the electric or magnetic fields on the closed surface S_c bounding V."

9.1.2   Scalar and vector potentials

The hard task of deriving expressions for $\mathbf{E}$ and $\mathbf{H}$ is enormously reduced if these fields are deduced from what are known as *potential functions*,[2] which have been in turn derived from charge and current distributions for a specific antenna. Due to the characteristics of the electric and magnetic fields, the former has associated a scalar potential, whereas the latter has a vector potential.

Provided that Equation (9.5) states that $\mathbf{B}$ must always be divergence free, i.e., $\nabla \cdot \mathbf{B} = 0$, this implies that it is possible to define $\mathbf{B}$ in terms of a vector potential, let us say $\mathbf{A}$, such that

$$\mathbf{B} = \mu\mathbf{H} \equiv \nabla \times \mathbf{A} \tag{9.8}$$

since the vector identity $\nabla \cdot (\nabla \times \mathbf{A}) \equiv 0$. Note that Equation (9.8) is given for any medium and therefore we are now using μ instead of μ_0. Thus, by substituting Equation (9.8) into Equation (9.2), it results in

$$\nabla \times \mathbf{E} + \frac{\partial}{\partial t}(\nabla \times \mathbf{A}) = 0$$

or equivalently,

---

[2]A potential is a mathematical function used to define the work done in any field.

$$\nabla \times \left(\mathbf{E} + \frac{\partial \mathbf{A}}{\partial t} \right) = 0 \tag{9.9}$$

Equation (9.9) implies that the curl of $(\mathbf{E} + \partial \mathbf{A}/\partial t)$ should be *zero* at all space points for all time, therefore this factor can be expressed in terms of a scalar potential Φ, which means [12]

$$\mathbf{E} + \frac{\partial \mathbf{A}}{\partial t} = -\nabla \Phi$$

or equivalently

$$\mathbf{E} = -\frac{\partial \mathbf{A}}{\partial t} - \nabla \Phi \tag{9.10}$$

9.1.3   Wave equations

As mentioned before, the relationship between electric and magnetic fields through Equations (9.2) and (9.3) provides one of the fundamentals of the electromagnetic theory. Indeed, the combination of these equations produces what is known as *wave equations* and it is therefore also applied to the potential functions introduced previously. Basically, from the substitution of Equations (9.8) and (9.10) into Equation (9.3) and using some vector identities, it is obtained [12] that

$$\nabla^2 \mathbf{A} - \frac{1}{c^2} \frac{\partial^2 \mathbf{A}}{\partial t^2} = -\mu \mathbf{J}_f + \nabla \left(\nabla \cdot \mathbf{A} + \frac{1}{c^2} \frac{\partial \Phi}{\partial t} \right) \tag{9.11}$$

where it is assumed that the medium within which the sources are embedded is linear, homogeneous, isotropic, and lossless. Thus, both its permittivity and permeability are constant parameters, ε, μ, and they are related to the speed of light through $c = 1/\sqrt{\varepsilon\mu}$.

In the same manner, by substituting Equation (9.10) into Equation (9.4), it results as

$$\nabla^2 \Phi + \frac{\partial}{\partial t} (\nabla \cdot \mathbf{A}) = -\frac{\rho_f}{\varepsilon} \tag{9.12}$$

For both (9.11) and (9.12), elements of free electric-charge and free electric-current have been assumed and hence the subscripts introduced in the charge and current densities (i.e., ρ_f and $\mathbf{J}_f$). From this pair of *coupled* second-order linear differential equations, Equation (9.11) can be simplified by taking into account that the curl and divergence operations of any vector are completely independent properties of the field. Thus, it is reasonable that

$$\nabla \cdot \mathbf{A} = -\frac{1}{c^2} \frac{\partial \Phi}{\partial t} \tag{9.13}$$

This expression is known as the *Lorentz Gauge*, from which, substituted into Equations (9.11) and (9.12) gives us the following *uncoupled*, linear, second-order differential wave equations:

$$\nabla^2 \mathbf{A} - \frac{1}{c^2}\frac{\partial^2 \mathbf{A}}{\partial t^2} = -\mu \mathbf{J}_f \tag{9.14}$$

$$\nabla^2 \Phi - \frac{1}{c^2}\frac{\partial^2 \Phi}{\partial t^2} = -\frac{\rho_f}{\varepsilon} \tag{9.15}$$

By analyzing the wave equations for only single-frequency solutions in a sinusoidal steady state and considering a harmonic source $e^{j\omega t}$, with ω as the angular frequency, it is possible to simplify their solution. Let us then represent the vector and scalar potentials respectively by

$$\mathbf{A}(x,\,y,\,z,\,t) = \Re\left[\underline{\mathbf{A}}(x,\,y,\,z)e^{j\omega t}\right] \tag{9.16}$$

$$\Phi(x,\,y,\,z,\,t) = \Re\left[\underline{\Phi}(x,\,y,\,z)e^{j\omega t}\right] \tag{9.17}$$

with $\underline{\mathbf{A}}(x,\,y,\,z)$ and $\underline{\Phi}(x,\,y,\,z)$ their complex amplitudes in sinusoidal steady state and $\Re(\cdot)$ the real part of the argument. Thus, by substituting Equations (9.16) and (9.17) into (9.14) and (9.15) for their time-independent form, it results as

$$\nabla^2 \underline{\mathbf{A}} = -\mu \underline{\mathbf{J}}_f \tag{9.18}$$

$$\nabla^2 \underline{\Phi} = -\frac{\underline{\rho}_f}{\varepsilon} \tag{9.19}$$

where $\underline{\mathbf{J}}_f$ and $\underline{\rho}_f$ are the complex amplitudes of $\mathbf{J}_f$ and ρ_f. These equations can be seen as the vector and scalar forms of the Poisson's equations, see Appendix C. Their solution for a certain but known distribution of $\underline{\mathbf{J}}_f$ and $\underline{\rho}_f$ at some observation point P is given by [12]

$$\underline{\mathbf{A}}_P = \frac{\mu}{4\pi}\int_V \frac{\underline{\mathbf{J}}_Q}{r_{QP}}dv_Q \tag{9.20}$$

$$\underline{\Phi}_P = \frac{1}{4\pi\varepsilon}\int_V \frac{\underline{\rho}_Q}{r_{QP}}dv_Q \tag{9.21}$$

where the subscript Q is for all possible source points and r_{QP} is the magnitude of the vector distance from a certain point Q to point P. In other words, it has been assumed that the point sources of $\mathbf{J}$ and ρ can be localized at *any* place within a finite volume V.

If the sources have a temporal dependence, the general solutions are [12]:

$$\mathbf{\underline{A}}_P = \frac{\mu}{4\pi} \int_V \frac{\mathbf{\underline{J}}_Q \, e^{-j\,\beta_0 r_{QP}}}{r_{QP}} dv_Q \qquad (9.22)$$

$$\underline{\Phi}_P = \frac{1}{4\pi\varepsilon} \int_V \frac{\underline{\rho}_Q \, e^{-j\,\beta_0 r_{QP}}}{r_{QP}} dv_Q \qquad (9.23)$$

with $\beta_0 = \omega\sqrt{\mu\varepsilon}$. Naturally, once $\mathbf{\underline{A}}_P$ and $\underline{\rho}_P$ are determined, the solution for **E** and **H** can be formulated.

9.1.4 Essence of the physically-sized antennas problem

Results of (9.22) and (9.23) have been derived for elemental electric and magnetic dipoles, which can be found in classical open literature. In any case, we are concerned with actual physically-sized antennas for which it can be stated that *"the total field produced by any antenna is then determinable (in principle, at least) as the vector sum of differential current-element electric-dipole contributions, integrated over the entire current distribution of the antenna."* [12]. Hence, it is clear that, mathematically and physically, the key point on the solution of the EM fields is the current distribution on the element under study, which is altered by boundary conditions. For this reason, it is important to address the meaning of both boundary value conditions and current distribution, as will be exposed in following sections.

9.2 Boundary Value Conditions

Now at this point, it is important to remark that the solution of the Maxwell's equations are subject to the boundary conditions imposed by the antenna geometry. In order to explain this concept, let us assume there is a certain structure (formed by a conductor or a combination of conductor and dielectric) of volume V in free space as depicted in Figure 9.1, where ε_r is the relative permittivity and μ_r is the relative permeability of the structure, which is also assumed to have a charge density ρ and a current density **J**.

We are interested in solving the relationship of the electric and magnetic fields not only at any point of free space due to the enclosed charge into the structure, but also at the boundary of it where there is a transition from one region (conductor/dielectric) to another one (free space). Then, let us illustrate a piece of the surface that encloses the volume V of Figure 9.1 as shown in Figure 9.2, which is a thin surface layer S between two regions R_1 and R_2.

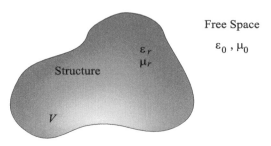

FIGURE 9.1
Arbitrary volumetric structure.

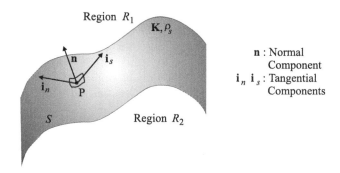

FIGURE 9.2
Thin surface layer.

This surface layer is characterized by having a surface charge density ρ_s and a surface current density $\mathbf{K}$, which introduce discontinuities in some field components.

9.2.1 Normal components

In the case of the normal components, there exists, on the one hand, a discontinuity in the component of $\varepsilon_0\mathbf{E}$ normal to the surface containing a layer of charge that is equal to ρ_s at this point. Mathematically,

$$\mathbf{n} \cdot \varepsilon_0 \left(\mathbf{E}_1 - \mathbf{E}_2\right) = \rho_s$$

where $\mathbf{E}_1$ and $\mathbf{E}_2$ are the electric field in regions R_1 and R_2 of Figure 9.2, respectively. On the other hand, the components of $\mu_0\mathbf{H}$ normal to the surface containing a surface charge and surface currents in free space are always continuous, which means

$$\mathbf{n} \cdot \mu_0 \left(\mathbf{H}_1 - \mathbf{H}_2\right) = 0$$

where $\mathbf{H}_1$ and $\mathbf{H}_2$ are the magnetic field in regions R_1 and R_2 of Figure 9.2, respectively.

9.2.2 Tangential components

Equivalently, there exist continuity and discontinuity on the tangential components of $\mathbf{E}$ and $\mathbf{H}$ at any particular surface point P. The continuity condition through the surface S is

$$\mathbf{n} \times (\mathbf{E}_1 - \mathbf{E}_2) = 0$$

Whereas the component of $\mathbf{H}$ tangent to S exhibits a discontinuity in passing through S equal to the component of $\mathbf{K}$ at right angles to it, in other words,

$$\mathbf{n} \times (\mathbf{H}_1 - \mathbf{H}_2) = \mathbf{K}$$

where $\mathbf{E}_1$, $\mathbf{E}_2$, $\mathbf{H}_1$ and $\mathbf{H}_2$ are as defined in Section 9.2.1 above.

9.2.3 Boundary conditions and unique solution to the Maxwell's equations

In summary, the boundary conditions relate the electric and magnetic fields both within a structure and to the surroundings and therefore they provide the additional expressions for the determining of the unique field solution of the system under study. These conditions are then

$$\mathbf{n} \times (\mathbf{E}_1 - \mathbf{E}_2) = 0 \tag{9.24}$$

$$\mathbf{n} \times (\mathbf{H}_1 - \mathbf{H}_2) = \mathbf{K} \tag{9.25}$$

$$\mathbf{n} \cdot \varepsilon_0 (\mathbf{E}_1 - \mathbf{E}_2) = \rho_s \tag{9.26}$$

$$\mathbf{n} \cdot \mu_0 (\mathbf{H}_1 - \mathbf{H}_2) = 0 \tag{9.27}$$

$$\mathbf{n} \cdot (\mathbf{J}_1 - \mathbf{J}_2) + \nabla_S \cdot \mathbf{K} = -\frac{\partial \rho_s}{\partial t} \tag{9.28}$$

9.2.4 Radiation condition

So far we have addressed boundary conditions between two or more media which can be a conductor, a dielectric, or a free space. Nevertheless, theoretically EM waves could travel unlimited in free space which is named an unbounded or open region, and therefore a boundary at infinity is mathematically stated. In order to have a unique solution for this problem, a condition must be specified which is known as a *radiation condition*. For this purpose, let us then assume that all sources and scattering objects embedded in free space are located at a finite distance from the origin of a coordinate system. Then, the electric and magnetic fields must satisfy

$$\lim_{r \to \infty} r \left[\nabla \times \begin{pmatrix} \mathbf{E} \\ \mathbf{H} \end{pmatrix} + jk_0\widehat{r} \times \begin{pmatrix} \mathbf{E} \\ \mathbf{H} \end{pmatrix} \right] = 0 \qquad (9.29)$$

which is known as *Sommerfeld's radiation condition*, with $r = \sqrt{x^2 + y^2 + z^2}$ and $k_0 = \omega\sqrt{\varepsilon_0\mu_0}$ the wavenumber in free space.

9.3   Current Distribution on Antennas

So far, the context of solving the Maxwell's equations for antennas has been explained and the importance of addressing the current distribution on them has been pointed out. As was stated in Chapter 2, an antenna can be seen as an open transmission line with some particular properties. This implies that an antenna is an element of a circuit such that depending on its geometric and electrical characteristics, certain capacitance, inductance, etc. are present through it and principles of circuit theory can be applied. Thus, basically the current distribution is a function that describes the form taken by the amplitude and phase of the current over its structure. This distribution depends on the materials, geometry, and dimensions of the antenna as well as its feed-point.

In order to illustrate the dependence on the antenna structure over the current distribution, Figures 9.3–9.7 show five basic antennas and its current distribution on each. All simulated antennas, except that of Figure 9.7, are classical narrowband and for all cases results were obtained with the CST Microwave Studio for a perfect electric conductor (a brief introduction of the software is given in Section 9.5 below). In all cases, a gray scale for the current distribution is presented, where the darkest tone represents the strongest intensity.

The first example, depicted in Figure 9.3, corresponds to the current distribution on an 1 GHz single wire monopole antenna. This antenna, mounted over a ground plane of 200 mm of diameter and 4 mm of thickness, has a length of 64.2 mm and a diameter of 2.4 mm. Its feed is the coaxial type and is located at the center of the ground plane whose substrate has a dielectric constant of 2.25. Due to the frequency of interest designed for this antenna, the simulation interval is from 0.5 to 2 GHz. The gray scale of the current distribution depicted in Figure 9.3 is for a range from 0 to 20.8 A/m approximately. As can be seen, the strongest intensity of current is concentrated near the ground plane at the feed-point.

The following example corresponds to a 1 GHz rectangular patch of 89.5 × 68.7 mm of wide and long, respectively and 0.3 mm thick, whose substrate of $\varepsilon_r = 4.6$ and thickness 3 mm is 156.6 mm wide and 179 mm long. The dimensions of the slot for the feed line is 1.86 mm wide and 27 mm long.

FIGURE 9.3
Current distribution for a 1 GHz wire antenna.

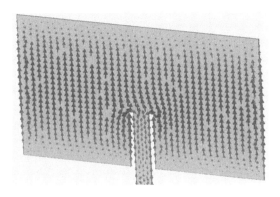

FIGURE 9.4
Current distribution for a 1 GHz rectangular patch.

In Figure 9.4 the current distribution for this antenna is shown, where the scale of values is from 0 to 45.8 A/m approximately. Due to the antenna characteristics, it is clearly appreciated a relatively more uniform distribution of the current in the middle area of the radiator, although, as expected the intensity fades toward the edges of the antenna.

A more complex structure is the following example, a log-periodic antenna of 7 elements. Its dimensions are summarized in Table 9.1, where all arms have a diameter of 10 mm and their lengths are taken from the center of the boom (which is 12 mm diameter, 692.64 mm long). Please note that the last column of this table refers to the separation between the i-th arm and the $(i-1)$-th arm, with i as the index of the arm. A scale factor from one element to the next one of 0.7 was used.

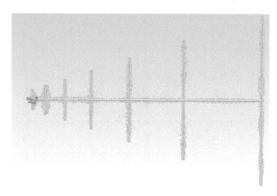

FIGURE 9.5
Current distribution for a 7-elements log-periodic antenna.

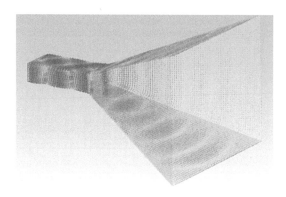

FIGURE 9.6
Current distribution for a 5 GHz horn antenna.

The simulations were carried out for a frequency interval from 0.3 to 1 GHz and whose results are shown in Figure 9.5 for 680 MHz, where the scale of the currents goes from 0 to 24.4 A/m approximately. Due to the geometry of the antenna, it is a little more complicated to appreciate the differences in the current distribution on this structure. However, more intensity can be seen near the shortest element, and there is a reduction of it on the largest element.

A 5 GHz horn antenna with 18 dBi of gain was considered as another example in this section, with the following dimensions: metal thickness of 0.6 mm; waveguide of 45.98 mm wide, 22.33 mm long with a longitude of 89.9 mm; distance from the horn aperture to the waveguide of 180.6 mm; horn aperture of 213.56 mm wide and 164.68 mm long. This antenna was simulated in a frequency interval from 2 to 8 GHz and its result for 5 GHz is shown in Figure 9.6. In this figure the scale of tones depicted comes from 0 to 2 A/m

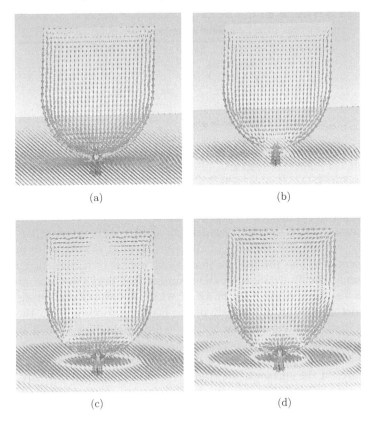

FIGURE 9.7
Current distribution for a planar monopole UWB antenna: (a) 3 GHz, (b) 7 GHz, (c) 11 GHz, (d) 15 GHz.

approximately. It is interesting to see how the wave propagation through the aperture of the antenna can be appreciated indirectly. The highest intensity of current is concentrated in the waveguide as the darkest tone shows in this figure.

The last example corresponds to a UWB planar monopole antenna operating from 2.6 to 15 GHz, whose dimensions are based on a circle of 21.112 mm in diameter. It is mounted 0.8 mm over a square ground plane of 84.44 mm side. The thickness of the metal is of 0.028 mm. This antenna is fed by a coaxial line whose central diameter is of 0.563 mm and its external diameter is of 1.32 mm. Its longitude to the ground plane is 1.68 mm. Due to the ultra wideband characteristics of this device, it was necessary to carry out simulations from 0 to 21 GHz. The results obtained are shown in Figure 9.7 for 3, 7, 11 and 15 GHz. A more uniform distribution of currents can be seen in the lowest bands of the antenna, whereas for higher frequencies, a reduction in the

TABLE 9.1

Dimensions of a 7-elements log-periodic antenna

i-th arm	Longitude (mm)	Separation to the $(i-1)$-th arm (mm)
1	375.0	–
2	262.5	235.5
3	183.75	164.85
4	128.62	115.39
5	90.03	80.77
6	63.02	56.54
7	44.11	39.58

current intensity is observed. As in other examples, the feed point presents a major concentration of current distribution.

For many years, distinct research has been conducted in order to determine the current distribution for diverse antennas [15–20] analytically and experimentally. The problem is that, as is well known, the current flow in an antenna usually depends on the resulting EM field generated around the antenna itself. There are a few cases for which the overall current distribution can be determined independently of the EM field solution (e.g., the physical electric-dipole antenna and the half-wave dipole antenna). When an analytical solution can be derived for some very specific geometries, they are called *canonical*, and are given for those boundary conditions for which one of the coordinates of the surface is constant in a coordinate system where the wave equation is separable [21].

There are also some relatively simple approximations to solutions of current distributions for more complex antenna structures like log-periodic antenna and parabolic reflector [20]. Moreover, an iterative methodology is suggested in [12] as a possible solution for an arbitrary shape antenna, but this possibility can always be subject to unrealistic results due to the trial-and-error approach.

In any case, it is clear that the starting point is to have the solution for a basic differential entity of the antenna (node, element, etc.) and then to solve the problem over the whole structure. The division of the antenna into basic entities and the posterior integration of the solution of each of them is the aim of the numerical methods.

9.4 Numerical Methods

9.4.1 Continuous equations in a finite computational environment

As can be seen from the set of Maxwell's equations (or the wave equation) given in Section 9.1, the unknown function is usually *continuous* and depends on *continuous* independent variables. That means that the set of partial differential equations used to explain the electromagnetic problem involves an infinite set of numbers. Of course, a certain manipulation to these equations should be applied in order to have a new set of equations with finite values suitable for a computer. This new set of equations results are algebraic and, in electromagnetics, usually linear. Thus, in order to find a solution for the set of linear algebraic equations, three general steps can be enumerated [6]:

1. Preprocessing. This step aims to derive the coefficients in the algebraic equations.

2. Solution of the algebraic equations.

3. Interpretation of the results.

As can be inferred, the form of how the structure of interest is taken to be solved in a computer and how transforming the EM problem given in an infinite domain to a finite domain (and its posterior solution) is the core of each numerical method. This is accomplished by different means, like projections through a linear combination of a certain finite set of basis functions onto the original domain, approximation of the field in a differential entity by polynomials, replacing the partial differential equations by algebraic expressions given for a function of the discrete index, and so on.

9.4.2   Computational domain and meshing

As has been expounded, a first question on the solution of the current distribution of practically any antenna, is to define the antenna characteristics, including its geometry, which naturally provides its boundaries. The definition of this structure in a computational device is what is known as a *computational domain* and corresponds to the *region where the fields are to be determined*. Once the computational domain has been specified, it has to be divided into basic entities over which the current is calculated depending on its own characteristics and as a function of its interaction with the neighboring entities. The subdivision of the domain into discrete entities is referred to as the *meshing* or *discretization* of the geometry of the antenna structure [8]. A second question derived from the first one is the size or separation between entities (i.e., the resolution), which, of course, will impact both the accuracy of results and the computing speed. A value of $\lambda/10$ or less per side is suggested in [8] if we are using elements as entities.

9.4.3   Classification of the methods

In accordance with the domain where a certain numerical method runs, there can be two groups [21]:

- **Time domain methods (TD)**: These methods are based on the temporal formulation of the electromagnetic problem.

- **Frequency domain methods (FD)**: In this case, the solution of the electromagnetic equations is constrained to sinusoidal steady state.

Another classification depends on the type of electromagnetic equations, so from [21]:

- **Differential methods**: Through what is known as the *finite difference* approach, the Maxwell's equations (or the wave equation) are discretized in such a way that a system of algebraic equations results. Depending on the domain where the method is implemented, it can be of *Finite Differences in the Time Domain* (FDTD), or *Finite Differences in the Frequency Domain* (FDFD).

- **Integral methods**: These methods are based on the application of the *Equivalence Theorem*[3] for the derivation of the field by means of the radiation integral of equivalent currents constrained to the boundary conditions. Depending on the techniques used to numerically impose the boundary conditions, the *Method of Moments* and the *Method of the Conjugated Gradient* can be distinguished.

- **Variational methods**: The variational methods are supported by the concept of a stationary functional[4] and that certain functions in turn make the functional achieve a maximum or minimum value. In the particular case of EM problems, the functional is a function of the electric and magnetic fields that present a maximum or minimum when the fields are precisely the solution of the problem. The functional is transformed in an equations system discretizing the fields and making zero the derivative of the functional respect to the parameters of the unknowns. The earliest variational technique is the Rayleigh–Ritz method, based on a very simple discretization of the solution. Currently the most popular variational method is known as the *Finite Element Method*, which approximates simultaneously the solution and the geometry by means of interpolation polynomials.

- **High frequency methods**: These methods are popular for those antennas whose structures are electrically large, like reflectors. It is due to the

---

[3]The Equivalence Theorem states that there exist equivalent currents (not real) which radiate in free space in the same field as in the real problem.

[4]A functional is a general form of a function, such that a functional associates a numerical value with each function. For example, let it be the length of curve bounded between (a, b) defined by a function $f(x)$ given by

$$L\left[f(x)\right] = \int_a^b \sqrt{1 + \left(\frac{df}{dx}\right)^2}\, dx$$

This is a special case of the functional

$$I[f] = \int_a^b F\left(x,\, f,\, \frac{df}{dx}\right) dx.$$

It is said that the functional is stationary when its variations are zero and it is neither maximum nor minimum.

need of discretizing the surface volume in elements of size of $\lambda/10$ or less that produces an equation system of elevated order, which requires larger computational resources. In order to solve this limitation, the Equivalence Theorem is applied in such a way that equivalent currents are assumed only on those surfaces over which the tangential components of the electric and magnetic fields can be estimated or known. For the rest of the surfaces, where it is not possible to know the equivalent currents, a perfect electric conductor is considered, so that instead of determining the equivalent currents based on the presence of free space, it is done for a conductor surface. Thus, the fields reflected on the surfaces and the fields diffracted in the edges can be determined by high frequency approximations. In these approaches the diffraction is locally considered, in such a way that no interaction between points of the surface is produced. An example of these methods is the well known *Ray Tracing*.

Much discussion has been produced as a result of the benefits and disadvantages of each approach used to solve the Maxwell's equations for antennas, in particular, and for electromagnetics, in general. The parameters considered in those discussions embrace convergence, computational complexity, etc. In any case, it is difficult to adopt a method for all possible problems at hand. However, nowadays some of the most popular methods for analyzing antennas are the *Difference Finite Method*, the *Finite Element Method* and the *Method of Moments*.

9.4.4 Finite differences method

The method of finite differences was developed before Maxwell's formalization of the electromagnetic theory. Although the basis of this method is attributed to Taylor, Jacob Stirling is considered the real founder of the calculus of finite difference, who also applied it to different areas of knowledge [22]. Based on his ideas, more than one century later in 1860, Boole was a pioneer in presenting a treatise on the subject for academic purposes [23]. The first development of this method for electromagnetic purposes was presented by Yee [24] in 1966. Independently the application, the fundamental idea of this method is to substitute the partial differential equations for a system of algebraic equations represented by a set of finite differences.

Then, let $f(x)$ be a continuous function as depicted in Figure 9.8 where x is also a continuous variable. By taking certain discrete values of the function, f_1, f_2, f_3 at discrete values of x, let's say, $x_0 - h$, x_0, $x_0 + h$, respectively, it is possible to express the first order derivative at x_0 as

$$\left.\frac{df}{dx}\right|_{x_0} \approx \frac{f_3 - f_2}{h} \tag{9.30}$$

$$\left.\frac{df}{dx}\right|_{x_0} \approx \frac{f_2 - f_1}{h} \tag{9.31}$$

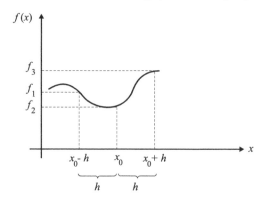

FIGURE 9.8
Example of discrete points from a one-dimensional continuous function.

$$\frac{df}{dx}\bigg|_{x_0} \approx \frac{f_3 - f_1}{2h} \tag{9.32}$$

In other words, we are approximating a continuous derivative at a point given as a *discrete difference* where h is the separation distance around the point under evaluation (in this case x_0). Difference equations given in (9.30), (9.31) and (9.32) are known as the *forward difference, backward difference* and *central difference*, respectively. From these, the central difference is used the most because it presents a smaller error [6,7,25]. By taking those intermediate points at a distance $h/2$ around x_0 and using the central difference, the following first order derivatives result:

$$\frac{df}{dx}\bigg|_{x_0+h/2} = \frac{f_3 - f_2}{h}$$

$$\frac{df}{dx}\bigg|_{x_0-h/2} = \frac{f_2 - f_1}{h}$$

Then, the second order derivative at x_0 can be expressed by,

$$\frac{d^2 f}{dx^2}\bigg|_{x_0} = \frac{\frac{df}{dx}\big|_{x_0+h/2} - \frac{df}{dx}\big|_{x_0-h/2}}{h} = \frac{f_3 - 2f_2 + f_1}{h^2}. \tag{9.33}$$

Let us now substitute f by Φ as a scalar potential as we had been previously using. If Φ is a function of two independent variables x and y, according to Figure 9.9, and by taking again the central difference as in (9.33) the second order derivative is

$$\nabla^2 \Phi = \frac{\partial^2 \Phi}{\partial x^2} + \frac{\partial^2 \Phi}{\partial y^2} = \frac{\Phi_1 + \Phi_2 + \Phi_3 + \Phi_4 - 4\Phi_0}{h^2} \tag{9.34}$$

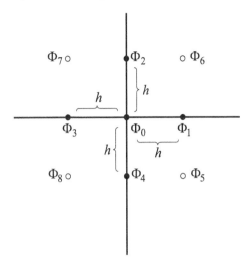

FIGURE 9.9
Array of nodes for a two-dimensional difference equation.

Please note that Figure 9.9 depicts five black points plus four white points. The idea is to represent a set of points, named *nodes*, around and including the node in question, in this case Φ_0. Although there can be several nodes that could be taken into consideration for the difference calculus, only those black nodes are considered, because they are at the same distance h from Φ_0. Equation (9.34) is therefore an equally spaced five-point difference equation. So, for the Poisson's equation (see Appendix C)

$$\frac{\Phi_1 + \Phi_2 + \Phi_3 + \Phi_4 - 4\Phi_0}{h^2} = -\frac{\rho}{\varepsilon} \qquad (9.35)$$

Then, a finite difference solution procedure is necessary for (9.35), which can be summarized as:

1. Define the computational domain or a suitable grid. That implies that the finite difference solution will provide discrete numbers of Φ at the nodes of the grid instead of using the continuous function $\Phi(x, y)$ for a given charge distribution $\rho(x, y)$.

2. Apply the difference relation given, in this case, by (9.33), at each node of the grid. The above produces N equations in N unknown node potentials.

3. Solve the resulting equations system. In order to do it, different classical approaches can be followed [26].

Naturally, some considerations should be taken into account, like the size of the separation distance, h; depending on the application, the possibility to

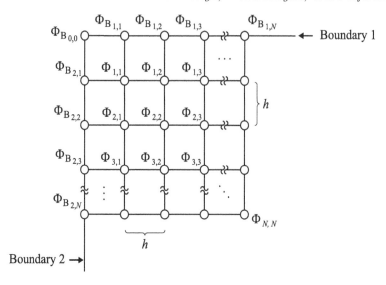

FIGURE 9.10
Boundary conditions for the finite difference method.

have a non-uniform grid; and the derivative boundary conditions [25]. From these, let us present some comments about the boundary conditions, taking Figure 9.10 as a reference. In this figure a rectangular grid is formed limited by two boundaries, where nodes on them have known values of Φ labeled as $\Phi_{B_{1,i}}$ and $\Phi_{B_{2,i}}$ with $i = 1, \ldots, N$ for boundary 1 and 2, respectively.

Let us consider first the Dirichlet boundary condition.[5] Let us transfer the five-point array of Figure 9.9 over the general grid of Figure 9.10 in such a way that the node corresponding to Φ_0 of Figure 9.9 coincides with node $\Phi_{1,1}$ of Figure 9.10. Under these conditions, values of nodes Φ_2 and Φ_3 of Figure 9.9 take the known values of Φ at boundaries 1 and 2, i.e., $\Phi_{B_{1,1}}$ and $\Phi_{B_{2,1}}$, respectively, whereas $\Phi_1 = \Phi_{1,2}$ and $\Phi_4 = \Phi_{2,1}$. Hence, the difference equation results as

$$\frac{\Phi_{1,2} + \Phi_{2,1} + \Phi_{B_{1,1}} + \Phi_{B_{2,1}} - 4\Phi_{1,1}}{h^2} = F(x_0, y_0)$$

where $F(x_0, y_0)$ is a known distribution evaluated at x_0, y_0.

There is also what is known as the *Neumann boundary condition* [6], where the value of the variable on the boundary is not known and its normal derivative is specified instead. In order to undertake this situation, a set of fictitious

---

[5]When the differential equation is subject to a known value of the variable in question (e.g., Φ) on the boundary of the domain of interest, it is known as the *Dirichlet boundary condition* [6].

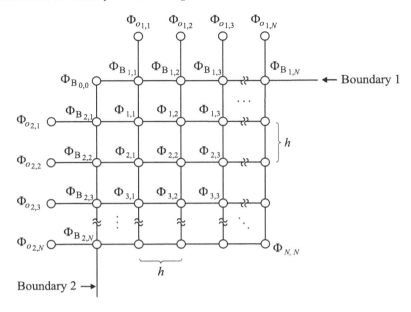

FIGURE 9.11
Fictitious nodes for the Neumann boundary condition.

nodes outside the boundaries should be considered as shown in Figure 9.11, which depicts the same grid of nodes as in Figure 9.10 but with some external fictitious nodes with values $\Phi_{o1,1}$, $\Phi_{o1,2}$, $\Phi_{o1,3}$, ..., $\Phi_{o1,N}$ for boundary 1 and $\Phi_{o2,1}$, $\Phi_{o2,2}$, $\Phi_{o2,3}$, ..., $\Phi_{o2,N}$ for boundary 2. Note that the subscript for all values of external nodes comes from outside.

Let us assume that the unknown values at boundary 1 have to be solved. By taking the five-point array as a basis of Figure 9.9 and the central node that of value $\Phi_{B1,1}$ of Figure 9.11, the difference equation results as

$$\frac{\Phi_{o1,1} + \Phi_{B0,0} + \Phi_{1,1} + \Phi_{B1,2} - 4\Phi_{B1,1}}{h^2} = F(x_0, y_0)$$

where the value of Φ on the corner is given by the average

$$\Phi_{B0,0} = \frac{\Phi_{B1,1} + \Phi_{B2,1}}{2}$$

As can be seen from the difference equation of this example, the external nodes are used to complete the five-point array on the boundary. Of course these values of fictitious nodes should not be solved for additional unknowns, but they have to be related to values inside the boundary. The above relation can be expressed through the central difference given in (9.32). Thus, for a node corresponding to $\Phi_{o1,1}$ for instance

$$\frac{\Phi_{o_{1,1}} - \Phi_{1,1}}{2h} = \left.\frac{\partial \Phi}{\partial x}\right|_{\Phi_{B_{1,1}}}$$

which is the specified value of the derivative on boundary 1. Hence an iterative method can be applied by means of the relation

$$\Phi_{o_{1,1}} = \Phi_{1,1} + 2h \left.\frac{\partial \Phi}{\partial x}\right|_{\Phi_{B_{1,1}}}$$

A detailed development of the application of this method for the Maxwell's equations can be found in the Yee's paper [24], where some simplifications have been assumed in order to compare numerical results with analytical approaches. In any case, the relative simplicity of the FD method has made it suitable for applications like microstrip transmission lines of different geometries [25] as well as waveguides and shielded parallel capacitance [6]. A free software for FD in time domain has even been developed, which is described in [27]. This method is straightforward for close regions where a finite number of points of the mesh or grid can be clearly defined. In contrast, for open regions there would be an infinite number of discrete points which obviously would be impractical to handle. In this case, a possible solution is to constraint the computational domain to a finite portion of space and then to successively expand it [6]. Naturally, care has to be taken with this approach due to the convergence of results.

Another particularity of the FD method is that it is based on a rectangular grid which provides a direct solution for rectangular boundaries. However, if the problem to solve has a curved boundary, a straight *stair-step* approach can be applied to it [6], where a set of nodes stair-step arranged outside and inside of the boundary can be separately considered and then their results can be averaged.

9.4.5   Finite element method (FEM)

As Huebner et al. tell briefly in [10], the background of the finite element method comes from the middle of the twentieth century when mathematicians, physics and engineers independently discovered techniques to solve numerically different problems like boundary value problems of continuous mechanics, and the stiffness influence coefficients of shell-type aerospace structures, among others. Perhaps one of the most popular problems is that of structural analysis where a certain structure is mathematically divided into a domain of continuous piecewise functions. Each of these functions is related to a non-overlapped subregion or cell, which has to be solved and then integrated into a whole entity. In any case, the essence of the FEM is exactly the form of the subdivision of the domain into small *elements* [8] and although the term element was used for several years, it was not until 1960 that Clough coined the name *finite element method* for this numerical procedure [28].

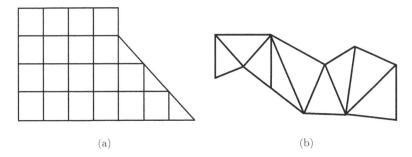

(a)   (b)

FIGURE 9.12
Examples of elements used to divide two-dimensional regions: (a) Combined triangle-square region, (b) irregular region into triangles.

9.4.5.1 Elements: The essence of the FEM

The general principle of the FEM states that the function to be evaluated in a region (probably large and complex) can be expressed by simple approximations in subregions (i.e., the computational domain), where these are named *finite elements*. For two-dimensional cases, these elements are polygons, typically triangles and quadrilaterals, which can be combined (see Figure 9.12a). Triangles are of particular interest due to their versatility of being like more irregular shapes as Figure 9.12b depicts. For three-dimensional structures, the elements can be tetrahedra, triangular prisms, or rectangular bricks, the tetrahedra being the simplest and best suited for arbitrary-volume shapes.

In comparison with the FD method previously presented, where a *pointwise* approximation is given to the equations that describe a certain problem, in the FEM a *piecewise* approximation is taken. This can be graphically illustrated in Figure 9.13, where the tulip-shaped monopole antenna explained in Chapter 3 is taken as an example. In general, one could expect more errors by using the FD method.

As can be observed, the FEM is based on the general operational principle of the numerical methods in terms of the discretization of the solution region or *continuum*[6]. For a more detailed description of guidelines to choosing the type of element, the interested reader can consult [10]. Now, independently of shape of elements, each of them is delimited by *edges*, which are in turn joined by *nodes*. In order to illustrate the above, we take as a basis the array of Figure 9.12b and marked elements, nodes, and some edges in Figure 9.14. As can be appreciated there are 9 triangular elements with 11 nodes, where the numbers of the last ones are contained in circles. In this figure, we pick out element 5 marking each edge of it under the label $\hat{e}_{ij}$, with i, j the numbers

---

[6] "A continuum is that body of matter (solid, liquid, or gas) or simply a region of space in which a particular phenomenon is occurring" [10].

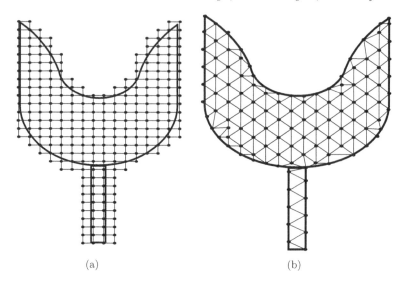

(a)   (b)

FIGURE 9.13
Graphical comparison of the division of the domain of the tulip-shaped
monopole antenna using (a) Pointwise (FD method) and (b) piecewise (FEM)
approximations.

of adjacent nodes (of course there are edges for each element, but we are only
showing them for one element to avoid overcrowding the figure).

On the other hand, as was mentioned in Section 9.4.3, the FEM corresponds to a variational method[7], what means that a functional (which is a
function of the electric and magnetic fields in the case of EM problems) is
defined instead of a function. Thus, the system of equations to solve is given
in terms not of field variables, but in terms of an integral-type functional.
This functional is then expressed over each element, where special care must
be taken when edges of adjacent elements are overlapped in such a way that
the field representations from one element to another conserve the continuity
of the field.

9.4.5.2   Interpolation functions

So far, we have briefly presented the first part of the FEM principle, that is
related to the division of the solution region into elements. Let us now explain some aspects associated with the representation of the function under
analysis through "simple" approximations. These, called *shape functions, basis functions* or *interpolation functions*, are defined in terms of values of the

---

[7]In addition to the variational principle, other approaches can be taken to express the
properties of individual elements, as will be explained below.

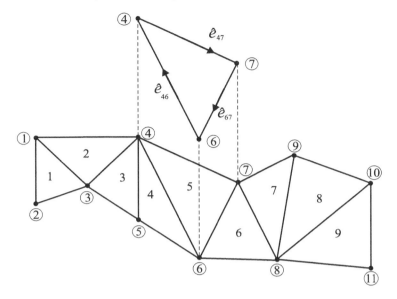

FIGURE 9.14
Nodes and edges in elements.

fields variables at specified points mentioned above as nodes and edges (sometimes called *nodal points* by some authors). Provided that there can be nodes both at the boundaries as at the center of the elements, the nodal values for the field variable plus the interpolation functions (let us name them together the *element model*) provide the necessary information of the field within the elements. The shape functions must have the following characteristics [8]:

- Spatial locality: As has been pointed out, basis functions are defined for each element in the FEM. Their influence is limited only to immediate neighboring elements.

- Approximation order: Polynomials are usually used as interpolation functions because they are easy to integrate and differentiate, where the grade will depend on the number of nodes of the element, the unknowns at each node, and the boundary conditions along the elements. Thus, the order of the approximation depends on the completeness of the polynomials. The general aim is to seek those expansion polynomials that will yield the highest order of approximation for a minimum number of unknowns associated with the element shape.

- Continuity: This term refers to the continuity of the derivatives, in such a way that functions with continuous derivatives up to the n-th order are said to be C^n continuous. Functions with C^0 continuity are used in most of the electromagnetic problems.

Nodal basis functions

The nodal basis functions are those local basis functions that are restricted to one element and are associated with each node of it. These functions can be denoted by $\varphi_i^e(x, y)$, where the superscript e stands for the number of element, whereas the subscript i corresponds to the local node number. Hence. the approximating function can be given by a linear combination of shape functions weighted by nodal coefficients, let us say, u_i^e. In other words, for a certain two-dimensional element, e, with p nodes,

$$\widetilde{\Phi}^e(x, y) = \sum_{i=1}^{p} u_i^e \varphi_i^e(x, y) \tag{9.36}$$

where Equation (9.36) must be valid for any u_i^e, so that $\varphi_i^e(x, y)$ must be unity at the i-th node and zero for all remaining nodes within the element:

$$\varphi_i^e(x_i^e, y_i^e) = 1, \quad \varphi_i^e(x_j^e, y_j^e) = 0 \quad \forall i \neq j \tag{9.37}$$

Edge basis functions

In contrast to node basis functions, the degrees of freedom of the edge basis functions are associated with the edges and faces of the finite element mesh. Let us consider a single two-dimensional element given by the rectangle shown in Figure 9.15, whose center is at (x_c^e, y_c^e) and is h_x^e long by h_y^e height. By assuming that each side of the rectangle is assigned a constant tangential field component, the field within the element can be expanded as [29]

$$E_x^e = \frac{1}{h_y^e} \left(y_c^e + \frac{h_y^e}{2} - y \right) E_{x1}^e + \frac{1}{h_y^e} \left(y - y_c^e + \frac{h_y^e}{2} \right) E_{x2}^e \tag{9.38}$$

$$E_y^e = \frac{1}{h_x^e} \left(x_c^e + \frac{h_x^e}{2} - x \right) E_{y1}^e + \frac{1}{h_x^e} \left(x - x_c^e + \frac{h_x^e}{2} \right) E_{x2}^e \tag{9.39}$$

where E_{x1}^e, and E_{x2}^e, represent the field components E_x along the edges $\hat{e}_{12}$, and $\hat{e}_{34}$, respectively whereas E_{y1}^e, and E_{y2}^e the corresponding components E_y along the edges $\hat{e}_{14}$, and $\hat{e}_{23}$, respectively. Now, the total field is given by the contribution of each tangential field component, denoted by E_i^e along the i-th edge. Thus, for the example of the rectangle it would be

$$\mathbf{E}^e = \sum_{i=1}^{4} \mathbf{N}_i^e E_i^e \tag{9.40}$$

where $\mathbf{N}_i^e$ are the vector basis functions given, respectively, for each edge, by

$$\mathbf{N}_1^e = \frac{1}{h_y^e} \left(y_c^e + \frac{h_y^e}{2} - y \right) \hat{x} . \tag{9.41}$$

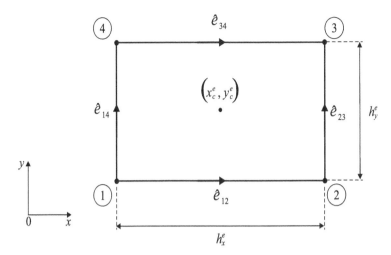

FIGURE 9.15
Edges on a rectangular element.

$$\mathbf{N}_2^e = \frac{1}{h_y^e} \left(y - y_c^e + \frac{h_y^e}{2} \right) \hat{x} \tag{9.42}$$

$$\mathbf{N}_3^e = \frac{1}{h_x^e} \left(x_c^e + \frac{h_x^e}{2} - x \right) \hat{y} \tag{9.43}$$

$$\mathbf{N}_4^e = \frac{1}{h_x^e} \left(x - x_c^e + \frac{h_x^e}{2} \right) \hat{y} \tag{9.44}$$

being edge 1, edge 2, edge 3, and edge 4 of those labeled $\hat{e}_{12}$, $\hat{e}_{34}$, $\hat{e}_{14}$, and $\hat{e}_{23}$ in Figure 9.15, respectively.

It is worth noting that these basis functions have a tangential component only along the i-th edge and none along all the other edges. This feature guarantees the continuity of the tangential field across all th element edges. Moreover, each basis function satisfies the divergence condition $\nabla \cdot \mathbf{N}_i^e$ within the element, hence they are suitable for representing the vector fields in source-free regions [29].

As is pointed out by [8,29] the nodal basis functions can be used for scalar nature problems like scalar potentials, mechanics, and so on. Nevertheless, they present some drawbacks for the solution of vector nature problems like those of electromagnetics and therefore the edge basis functions are more convenient. For example, nodal basis functions impose continuity in all three spatial components and in contrast, as we have just seen, edge basis functions guarantee continuity only along the tangential component. The above allows us to address problems with discontinuous material boundaries. On the other hand, special care must be taken when nodal basis functions are used for enforcing boundary conditions at material interfaces, conducting surfaces, and geometry corners [8].

9.4.5.3   Sequence of the finite element method

From the specification of the element model, the properties of the element can be stated through matrix equations. There are three ways to do the above [10]: the direct approach, the variational approach, and the weighted residuals approach. The term *direct approach* is simply attributed to the direct stiffness method of structural analysis for only relatively simple problems (simple element shapes). The *variational approach*, as we have pointed out, is based on a stationary functional, where certain functions make the functional achieve either a maximum or a minimum value. Finally, in contrast, when there is not a functional available, the *weighted residuals approach* takes the equations that describe the phenomenon under analysis and proceeds directly, avoiding the variational statement. In any case, the basic steps of the FEM can be summarized and sequentially applied as follows [10, 29]:

1. Discretization or subdivision of the domain or continuum
2. Selection of the interpolation functions
3. Determination of the element properties
4. Formulation of the system of equations
5. Solution of the system of equations

With the above structure in mind, some general guidelines can be taken from [30]. Let $\mathcal{L}\Phi = f$ be the differential equation to be solved, where $\mathcal{L}$ is the differential operator, f is the source or excitation function, and Φ is the unknown function to be determined in a certain region Ω. Then:

- The first step consists of proceeding to the subdivision of Ω into elements (although not in detail, some comments about elements have been stressed before).

- Continue with the approximation of the solution by an expansion in a finite number of basis functions in such a way that

$$\Phi \approx \sum_{i=1}^{n} \Phi_i g_i \tag{9.45}$$

where Φ_i are unknown coefficients multiplying the basis functions or interpolation functions g_i. Regarding selection of the interpolation functions, as mentioned before, polynomials are usually used as interpolation functions, where the grade will depend on the number of nodes of the element, the unknowns at each node and the boundary conditions along the elements (see [10, 29] for details about how to select an interpolation function).

- Once the elements and their interpolation functions have been selected, the element properties can be defined. Let us assume for the moment that the weighted residual approach is taken for this aim.

- The following step is to form the residual $r = \mathcal{L}\Phi - f$ which is expected to be as small as possible. In general, it will not be zero in a pointwise sense, but it will be in the so-called weak sense by setting a weighted average of it to zero. Then, *test* or *weighting* functions w_i with $i = 1, \ldots, n$ have to be chosen for weighting the residual r. The number n will depend on the unknown coefficients. When the weighting functions are the same as the basis functions, i.e., $w_i = g_i$, the procedure is known as *Galerkin's method*, which will be illustrated later.

- Finally, set the weighted residuals to zero and solve for the unknowns Φ_i, which means solve the set of equations

$$\langle w_i, r \rangle = \int_\Omega w_i \, r \, d\Omega = 0 \tag{9.46}$$

In order to illustrate the FEM numerically, let us take as a basis the approaches to determine the element properties. Starting with the variational approach through a comparison of the Rayleigh–Ritz method and the FEM; then we take the weighted residuals approach where some comments about the difference between the Galerkin's method and the FEM are given. Regarding the direct approach, we do not present an example here, but the interested reader can consult [10].

9.4.5.4 Rayleigh–Ritz method versus FEM

The method first developed by Rayleigh and later improved by Ritz [6] follows the first steps of the general guidelines explained before, where the function is approximated by a linear combination of known functions, as expressed by Equation (9.45), whose solution consists of determining the parameters in the combination. However, taking the variational principle to determine the element properties, that lineal combination is substituted into a functional in order to determine if it is minimum or maximum through the differentiation with respect to each parameter. Then, let us consider, for instance, the following functional

$$I(f) = \int_0^1 \left[\left(\frac{df(x)}{dx} \right)^2 + f^2(x) \right] dx \tag{9.47}$$

with $f(0) = 0$, $f(1) = 1$. A first function that satisfies the boundary conditions without minimizing the functional could be

$$f(x) = x \tag{9.48}$$

Then, a set of functions given by

$$\sum_{j=1}^{N} \alpha_j \left(x - x^{j+1}\right) \tag{9.49}$$

could be added to improve the first approach (9.48) in such a way that

$$f(x) = x + \sum_{j=1}^{N} \alpha_j \left(x - x^{j+1}\right) \tag{9.50}$$

with α_j parameters to be determined. Equation (9.50) can be expressed in a general notation as

$$f(x) = f_0(x) + \sum_{j=1}^{N} \alpha_j f_j(x) \tag{9.51}$$

where $f_j(x)$ satisfies the homogeneous boundary conditions

$$f_j(0) = 0, \ f_j(1) = 0$$

Substitution of (9.51) into (9.47) and after expanding binomials and grouping terms, yields:

$$
\begin{aligned}
I(f) = &\int_0^1 \left[\left(\frac{df_0(x)}{dx}\right)^2 + f_0^2(x) \right] dx + \\
&+ 2\alpha_j \sum_{j=1}^{N} \int_0^1 \left[\left(\frac{df_0(x)}{dx}\right) \left(\frac{df_j(x)}{dx}\right) + f_0(x) f_j(x) \right] dx + \\
&+ \sum_{j=1}^{N} \sum_{k=1}^{N} \alpha_j \alpha_k \int_0^1 \left[\left(\frac{df_j(x)}{dx}\right) \left(\frac{df_k(x)}{dx}\right) + f_j(x) f_k(x) \right] dx
\end{aligned}
\tag{9.52}
$$

whose differentiation with respect to each α_j results as [6]

$$
\begin{aligned}
&\sum_{k=1}^{N} \alpha_k \int_0^1 \left[\left(\frac{df_j(x)}{dx}\right) \left(\frac{df_k(x)}{dx}\right) + f_j(x) f_k(x) \right] dx \\
&= -\int_0^1 \left[\left(\frac{df_0(x)}{dx}\right) \left(\frac{df_j(x)}{dx}\right) + f_0(x) f_j(x) \right] dx
\end{aligned}
\tag{9.53}
$$

which are set to zero in order to have N linear algebraic equations. The solution of this set of equations corresponds to the parameters α_k and therefore the

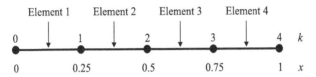

FIGURE 9.16
Division of the interval $(0, 1)$ into four equally spaced subintervals.

approximate solution for the function that makes the functional stationary, the principle of the variational methods.

Let us now take an example as is presented in [6] to illustrate the FEM in a one dimensional region over which the functional given in (9.47) has to be minimum. By dividing the interval $(0, 1)$ into four one dimensional subintervals (or elements) as Figure 9.16 shows which can be expressed as

$$f_k = f(0.25k)$$

Then, let us assume the following function that approximates to the functional as

$$f(x) = f_{k-1} + 4\left(f_k - f_{k-1}\right)\left[x - \frac{1}{4}(k-1)\right]$$

which substituted in Equation (9.47) yields

$$I(f) = \int_0^1 \left(\frac{d}{dx}\left[f_{k-1} + 4\left(f_k - f_{k-1}\right)\left[x - \frac{1}{4}(k-1)\right]\right]\right)^2 dx +$$

$$+ \int_0^1 \left(f_{k-1} + 4\left(f_k - f_{k-1}\right)\left[x - \frac{1}{4}(k-1)\right]\right)^2 dx$$

$$= 16\left(f_k - f_{k-1}\right)^2 \int_0^1 dx + f_{k-1}^2 \int_0^1 dx +$$

$$+ 8f_{k-1}\left(f_k - f_{k-1}\right)\int_0^1 \left[x - \frac{1}{4}(k-1)\right] dx +$$

$$+ 16\left(f_k - f_{k-1}\right)^2 \int_0^1 \left[x - \frac{1}{4}(k-1)\right]^2 dx$$

Then the evaluation of the integrals over each of subintervals yields

$$I(f) = 4f_1^2 + 4\left(f_2 - f_1\right)^2 + 4\left(f_3 - f_2\right)^2 + 4\left(1 - f_3\right)^2 +$$

$$+ \frac{1}{12}\left(2f_1^2 + f_1 f_2 + 2f_2^2 + f_2 f_3 + 2f_3^2 + f_3 + 1\right)$$

$$= \frac{49}{6}\left(f_1^2 + f_2^2 + f_3^2\right) - \frac{95}{12}\left(f_1 f_2 + f_2 f_3 + f_3\right) + \frac{49}{12}$$

Setting the derivatives of $I(f)$ with respect to f_1, f_2, and f_3 equal to zero, the following set of algebraic equations is obtained

$$\frac{49}{3} f_1 - \frac{95}{12} f_2 = 0$$

$$\frac{49}{3} f_2 - \frac{95}{12} (f_1 + f_3) = 0$$

$$\frac{49}{3} f_3 - \frac{95}{12} (f_2 + 1) = 0$$

whose solutions are approximately

$$f_1 = 0.21$$

$$f_2 = 0.44$$

$$f_2 = 0.70$$

which approximate well to the analytic solution. As can be appreciated, both the Rayleigh–Ritz method and the FEM (in its variational approach) share the same form of closing to the solution through a linear combination of certain functions to make a functional stationary. However, its major difference is that for the FEM these functions are not defined in a whole region, but in subintervals, the essence of the method. Hence, while the Rayleigh–Ritz technique is a global-domain method, the FEM is applied to discrete local domains, which are usually simpler than that of a global domain. Therefore, the FEM can be considered as a special case of the Rayleigh–Ritz method.

9.4.5.5 FEM in the context of the Galerkin's method

As was already mentioned, another approach that can be taken within the FEM to determine the element properties is that based on the weighted residual techniques. An example of this type of method is that developed by Galerkin.

Let be $\tilde{\Phi}$ the approximate solution of a differential equation expressed by

$$\mathcal{L}\Phi = f \tag{9.54}$$

where, as stated before, $\mathcal{L}$ is the differential operator, f is the source or excitation function, and Φ is the unknown function to be determined in a certain region Ω. By substituting $\tilde{\Phi}$ for Φ in (9.54) the following residual results:

$$r = \mathcal{L}\tilde{\Phi} - f \tag{9.55}$$

which could not be necessarily equal to zero. Then a good approximation for $\tilde{\Phi}$ would be that where (9.55) is reduced to the least value at all points of region Ω. This can be achieved by

$$R_i = \int_\Omega w_i \, r \, d\Omega = 0 \tag{9.56}$$

where R_i denotes the weighted residual integrals and w_i the weighting functions. Let us assume that $\widetilde{\Phi}$ can be approximated as in (9.45) by

$$\widetilde{\Phi} = \sum_{i=1}^{n} \Phi_i \, g_i \tag{9.57}$$

for which Φ_i represents the constant unknown coefficients and g_i the chosen interpolation functions. Equation (9.57) can be also expressed by

$$\widetilde{\Phi} = \{\Phi\}^T \{g\} = \{g\}^T \{\Phi\} \tag{9.58}$$

where $\{\cdot\}$ denotes a column vector and the superscript T denotes the transpose of the vector.

By substituting (9.58) into (9.55) and it in turn into (9.56) for $w_i = g_i$ (corresponding to the Galerkin's method) and for which the most accurate solution is found [29], it yields

$$R_i = \int_{\Omega} g_i \left[\mathcal{L} \{\Phi\}^T \{g\} - f \right] d\Omega = 0 \tag{9.59}$$

Let us follow a one dimensional example as described in [29]. Let us take the Poisson's equation given in Appendix C, which can be used to describe the static potential Φ between two infinite parallel plates, where one of them is located at $x = 0$ with $\Phi = 0\,\text{V}$ and the other one is at $x = 1\,\text{m}$ with $\Phi = 1\,\text{V}$ for which a constant ε can be considered between plates and a varying electric charge density is given by

$$\rho(x) = -(x+1)\varepsilon \tag{9.60}$$

Then, the Poisson's equation is

$$\frac{d^2\Phi}{dx^2} = x + 1 \quad ; \quad 0 < x < 1 \tag{9.61}$$

with boundary conditions

$$\Phi|_{x=0} = 0 \tag{9.62}$$

$$\Phi|_{x=1} = 1 \tag{9.63}$$

Then, let us apply the weighted residual method to the problem defined by Equations (9.61)–(9.63), whose exact solution is

$$\Phi(x) = \frac{1}{6}x^3 + \frac{1}{2}x^2 + \frac{1}{3}x \tag{9.64}$$

First, let us take the residual given in (9.55) to (9.61), where we are also substituting an approximated $\widetilde{\Phi}$ by Φ

$$r = \frac{d^2\widetilde{\Phi}}{dx^2} - x - 1 \tag{9.65}$$

from where the weighted residual integrals are

$$R_i = \int_0^1 w_i \left(\frac{d^2 \tilde{\Phi}}{dx^2} - x - 1 \right) = 0 \qquad (9.66)$$

Let us assume an expansion of $\tilde{\Phi}$ in terms of polynomials of the form

$$\tilde{\Phi}(x) = c_1 + c_2 x + c_3 x^2 + c_4 x^3 \qquad (9.67)$$

and given the boundary conditions (9.62) and (9.63), from which $c_1 = 0$ and $c_2 = 1 - c_3 - c_4$, the expansion (9.67) is simplified to a two coefficients expression

$$\tilde{\Phi}(x) = x + c_3(x^2 - x) + c_4(x^3 - x) \qquad (9.68)$$

Thus, we need two weight functions which are taken as $w_1 = x^2 - x$ and $w_2 = x^3 - x$ according to Galerkin's method. Hence, using these weight functions and applying (9.68) to (9.66), a system of two equations yields:

$$\frac{1}{2}c_4 + \frac{1}{3}c_3 - \frac{1}{4} = 0 \qquad (9.69)$$

$$\frac{4}{5}c_4 + \frac{1}{2}c_3 - \frac{23}{60} = 0 \qquad (9.70)$$

whose solution results are $c_3 = 1/2$ and $c_4 = 1/6$, which substituted in (9.68) coincide with the exact solution given in (9.64). The concordance between analytical and numerical results obeys the simple nature of the problem, which also allows to have a complete basis; otherwise, only an approximate solutions will be found. Moreover, as has been pointed out from the beginning of the chapter, an analytic solution is not always possible to have, which depends on the complexity of the problem.

Another aspect illustrated in this example is the importance of defining the trial functions as a step in Galerkin's method, which are specified over the entire solution domain. When no direct trial functions can be expressed over the whole domain, it is convenient to subdivide the entire domain into small subdomains and use trial functions defined over each of them. This is, again, the core of the FEM. For the example of the parallel infinite plates described by Equations (9.61)–(9.63), in the FEM method the domain is then divided. Let us consider, for instance, a uniform division of the one dimensional domain $(0, 1)$ into three subdomains defined by (x_1, x_2), (x_2, x_3), and (x_3, x_4) with $x_1 = 0$ and $x_4 = 1$ according to the boundary conditions, and where the other two points can be located at $x_2 = 1/3$ and $x_3 = 2/3$. A linear variation of $\Phi(x)$ over each interval defined by

$$\tilde{\Phi}(x) = \Phi_i \frac{x_{i+1} - x}{x_{i+1} - x_i} + \Phi_{i+1} \frac{x - x_i}{x_{i+1} - x_i} \qquad (9.71)$$

can be assumed as an expansion function for $x_i \leq x \leq x_{i+1}$ with $i = 1, 2, 3$. By inspecting (9.71), it is clear that the unknown constants Φ_i represent the value of $\Phi(x)$ at $x = x_i$. Moreover, the boundary conditions of the problem imply that $\Phi_1 = 0$ and $\Phi_4 = 1$, and therefore only Φ_2 and Φ_3 have to be obtained. By following the principle of Galerkin's method, the weighting functions are chosen as the expansion function given in (9.71), i.e.,

$$
w_i = \begin{cases} \frac{x - x_{i-1}}{x_i - x_{i-1}}, & x_{i-1} \leq x < x_i \\[2mm] \frac{x_{i+1} - x}{x_{i+1} - x_i}, & x_i \leq x \leq x_{i+1} \end{cases} \tag{9.72}
$$

Now, by observing the approximation given in Equation (9.71), it can be differentiated only once, hence it is necessary to reduce the order of the differentiation in order to substitute $\widetilde{\Phi}$ into (9.66). The above can be done by integration by parts, which results in the transfer of one derivative to the weighting function, in such a way that,

$$
\int_{x_{i-1}}^{x_{i+1}} w_i \left(\frac{d^2 \widetilde{\Phi}}{dx^2} \right) dx = w_i \left. \frac{d\widetilde{\Phi}}{dx} \right|_{x_{i-1}}^{x_{i+1}} - \int_{x_{i-1}}^{x_{i+1}} \frac{dw_i}{dx} \frac{d\widetilde{\Phi}}{dx} dx \tag{9.73}
$$

By using this result in (9.66) and since w_i vanishes at x_{i-1} and x_{i+1}, the residual integrals result as

$$
\int_{x_{i-1}}^{x_{i+1}} \frac{dw_i}{dx} \frac{d\widetilde{\Phi}}{dx} dx + \int_{x_{i-1}}^{x_{i+1}} (x+1) w_i dx = 0 \tag{9.74}
$$

From where, by substituting $\widetilde{\Phi}$ as in (9.71) and the weighting functions given in (9.72) into Equation (9.74), the following system of two equations is obtained

$$
6\Phi_2 - 3\Phi_3 + \frac{4}{9} = 0 \tag{9.75}
$$

$$
-3\Phi_2 + 6\Phi_3 + \frac{22}{9} = 0 \tag{9.76}
$$

whose solution is $\Phi_2 = 14/81$ and $\Phi_3 = 40/81$. Then, with this solution we have all the values of x_i (remember that x_1 and x_4 were determined from the boundary conditions) and $\Phi(x)$ at any x can be obtained simply by interpolation using (9.71). Small differences are presented between this numerical solution and the exact solution given in (9.64) mainly in those points different from the exact values of x_i [29].

Regarding the FEM in the context of the weighted residual techniques, as in the variational Ritz–Rayleigh method, both solutions are practically comparable for simple one-dimensional cases described, being the division of the domain and therefore the definition of the trial functions over each subdomain in the FEM the difference. As was mentioned, this division of the domain becomes of special importance for two and three dimensional structures or with high irregular boundaries where the definition of the trial functions can become quite difficult or impossible.

9.4.6 Method of Moments (MoM)

9.4.6.1 Introduction

It is worth introducing the final method discussed in this chapter with some historical notes taken from [31], where it is asserted that the MoM is more of a concept than a method. Indeed, this corresponds to the field of the transforming of a linear functional equation into a linear matrix equation (i.e., projections from an infinite dimensional function space onto a finite dimensional subspace), whose mathematical basis was stated by Hilbert and used later in quantum mechanics in the 1920s. Some of the early references to applications of the MoM published during 1930s can be found in the chapter *The general theory of approximation methods* by Kantorovich and Akilov [32]. The development of computers during World War II opened the possibility of implementing solutions emerged from the MoM to cumbersome problems like resolution of a huge linear system of equations, matrix inversion, and so on. Thus, in 1960 it was possible to apply the exploration of the MoM to the solution of electromagnetics problems. The early works of application of the MoM to electromagnetics are attributed to Harrington, who first presented a paper [33] and later a book on the matter [34]. Some few years later the popularity of the method was evident even for undergraduate courses [35] and text books like [7].

9.4.6.2 Mathematical principle

In essence, an integro-differential equation for a given problem (such as those described by Maxwell's equations) can be represented as an infinite-dimensional functional equation. For example,

$$\mathcal{L}f = g \tag{9.77}$$

where $\mathcal{L}$ is the linear operator associated with the integro-differential equation, g is a known function related to the source (e.g., the incident field), and f is an unknown function to be determined (e.g., an induced current distribution). In order to achieve the projection into a finite dimensional subspace, f is represented by a linear combination of a finite set of basis functions, f_j. In other words, f is expanded in a series of functions f_1, f_2, f_3, ... in the domain of $\mathcal{L}$ as [8]

---

[8]The domain of $\mathcal{L}$ represents the functions f on which it operates.

$$f \approx \sum_j \alpha_j f_j \tag{9.78}$$

and therefore f_j are also known as *expansion functions*. In (9.78) α_j are discrete samples of f and therefore represent the unknowns of the numerical problem to solve. In order for (9.78) to be a good approximation, the basis functions f_j have to be in the domain of $\mathcal{L}$, so that the differentiation and boundary conditions of the operator are fulfilled.

By substituting (9.78) into (9.77), and using the linearity of $\mathcal{L}$

$$\sum_j \alpha_j \mathcal{L} f_j \approx g \tag{9.79}$$

Let us define a finite set of *weighting functions* or *testing functions*, w_j, in the range of $\mathcal{L}$, where the range corresponds to the functions g resulting from the operation. Through the operation of the inner product (usually an integration) applied to the functional expansion with each weighting function, and using its linearity, it is possible to obtain a finite set of equations for the coefficients of the basis functions. Let us take $\langle f, g \rangle$ as the inner product suitable for the problem; then by applying it to (9.79)

$$\sum_j \alpha_j \langle w_i, \mathcal{L} f_j \rangle = \langle w_i, g \rangle \tag{9.80}$$

with $i = 1, 2, 3, \ldots$ This set of equations is then solved to obtain the solution for f, which is exact if the sum in (9.78) is infinite and forms a complete set of basis functions, or approximate if it is finite. The set of equations given in (9.80) can be expressed in matrix form as

$$\mathbf{L}_{ij}\alpha_j = \mathbf{g}_i \tag{9.81}$$

where

$$\mathbf{L}_{ij} = \begin{bmatrix} \langle w_1, \mathcal{L} f_1 \rangle & \langle w_1, \mathcal{L} f_2 \rangle & \cdots \\ \langle w_2, \mathcal{L} f_1 \rangle & \langle w_2, \mathcal{L} f_2 \rangle & \cdots \\ \vdots & \vdots & \ddots \end{bmatrix} \tag{9.82}$$

$$\alpha_j = \begin{bmatrix} \alpha_1 \\ \alpha_2 \\ \vdots \end{bmatrix} \tag{9.83}$$

$$\mathbf{g}_i = \begin{bmatrix} \langle w_1, g \rangle \\ \langle w_2, g \rangle \\ \vdots \end{bmatrix} \tag{9.84}$$

If the matrix $\mathbf{L}$ is nonsingular, its inverse $\mathbf{L}^{-1}$ exists. The α_j are then given by

$$\alpha_j = \mathbf{L}_{ij}^{-1}\mathbf{g}_i \tag{9.85}$$

and the solution for f is given by Equation (9.78).

Now, let us take a finite approach of Equation (9.78), let us say f_N,

$$f_N = \sum_j^N \alpha_j f_j \tag{9.86}$$

from which,

$$g_N = \sum_j^N \alpha_j \mathcal{L} f_j \tag{9.87}$$

Then, the error in the boundary conditions or residual is

$$R = g - g_N = g - \sum_j^N \alpha_j \mathcal{L} f_j \tag{9.88}$$

9.4.6.3   Basis functions to discrete 3D arbitrary surfaces

Sometimes the EM analysis of antennas has to be carried out in the presence of other objects, like those antennas embedded in airplanes, ships or satellites, which produce quite complex structures to simulate. Nowadays there exists commercial software specialized to discrete arbitrary surfaces and to form the mesh over which the analysis will be done. These programs are based on basis functions which are briefly described as follows [21].

1. Wire-grid modeling: This is the oldest technique, developed in the middle of the 1960s. Basically the radiation of superficial equivalent currents is approximated by the radiation of wires of currents over the surface of the object under analysis. This technique has good enough results to determine far field parameters; however, it is not suitable for near field parameters, which seriously limits this approach. Figure 9.17 shows the conical antenna with a circular aperture used as a basis to model a planar directional UWB antenna as discussed in Chapter 7, and taken here as an example to illustrate the wire-grid modeling.

2. Triangles of Rao, Wilton and Glisson: Rao, Wilton and Glisson developed in 1982 a modeling technique based on triangle patches [36], for which a special type of basis function is defined over the

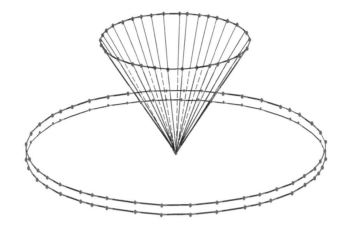

FIGURE 9.17
Modeling of surface by a wire-grid of current.

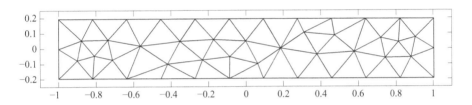

FIGURE 9.18
A surface mesh of 74 triangles.

mesh formed. In other words, the idea is to construct an arbitrary surface using triangles as shown in Figure 9.18.

In order to avoid singularities of charge in a mesh of triangular patch, it is important to preserve the continuity of the component **J** normal to the triangle sides, assuming that the basis functions are continued inside them. Thus, the Rao, Wilton and Glisson (RWG) basis functions are defined as vector functions $\mathbf{f}_n$ associated to one side ℓ_n that is common to a couple of triangles T_n^+ and T_n^- (see Figure 9.19)

$$
\mathbf{f}_n(\mathbf{r}) = \begin{cases} \dfrac{\ell_n}{2A_n^+}\rho_n^+, & \mathbf{r} \in T_n^+ \\[2ex] \dfrac{\ell_n}{2A_n^-}\rho_n^-, & \mathbf{r} \in T_n^- \\[2ex] 0 & \text{otherwise} \end{cases} \tag{9.89}
$$

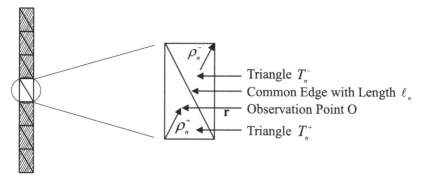

Strip divided in
right triangles

FIGURE 9.19
Edge element on a strip line.

where $\mathbf{r}$ is a vector from an observation point O to either triangle $T_n^{\pm}$, $\rho^{\pm}$ are position vectors defined with respect to the free vertex of $T_n^{\pm}$. Note that whereas ρ^+ comes from the free vertex of T_n^+ to $\mathbf{r}$, ρ^- is directed from $\mathbf{r}$ toward the free vertex of T_n^-. This is so, according to the convention originally presented by Rao et al. [36] with respect to the choice of a positive current reference direction (from T_n^+ to T_n^-). The properties that result from definition (9.89) can be found in [36], from which it is derived that the total current on the surface of the structure made up by N edge elements (i.e., each pair of triangles having a common edge) is

$$\mathbf{J} \cong \sum_{n=1}^{N} I_n \mathbf{f}_n(\mathbf{r}) \tag{9.90}$$

where coefficients I_n are the current values that pass through the side ℓ_n in the normal direction.

3. Finite elements: Several comments and references were presented in the previous section related to finite elements. Therefore no more details will be considered here.

9.4.6.4 MoM steps

The steps of the MoM can be stated in a few words, as Harrington summarizes in [37]:

- "Define an inner product appropriate to the problem.

- Choose a set of expansion functions in terms of which the unknown is approximated by a linear combination.

- Choose a set of testing functions to define a subspace in which the approximate solution is to be valid.

- Set the inner products of the approximate known equal to the corresponding inner product of the exact known for each testing function.

- The result is a set of linear equations to determine the coefficients of the approximate solution."

9.4.6.5 Example

In this section let us take a simple example as illustrated in [34]. By considering the problem

$$-\frac{d^2 f}{dx^2} = 1 + 4x^2 \tag{9.91}$$

with boundary conditions

$$f(0) = f(1) = 0 \tag{9.92}$$

In this problem the specific source has been obviously taken as $g = 1 + 4x^2$ and has as analytic solution

$$f(x) = \frac{5x}{6} - \frac{x^2}{2} - \frac{x^4}{3} \tag{9.93}$$

Let us then determine the numerical solution by the MoM. First, by choosing a finite power-series solution as

$$f_j = x - x^{j+1} \tag{9.94}$$

where $j = 1, 2, \ldots, N$ in such a way that (9.78) is

$$f = \sum_{j=1}^{N} \alpha_j \left(x - x^{j+1} \right) \tag{9.95}$$

Let us assume that the Galerkin's method is adopted in such a way that

$$w_j = f_j = x - x^{j+1} \tag{9.96}$$

A suitable inner product for this problem is

$$\langle f, g \rangle = \int_0^1 f(x)g(x)dx \tag{9.97}$$

Thus, by applying (9.97) into matrices (9.82)–(9.84) and considering that $\mathcal{L} = -d^2/dx^2$, results in

$$L_{ij} = \langle w_i, \mathcal{L}f_j \rangle = \frac{ij}{i + j + 1} \tag{9.98}$$

TABLE 9.2
Solution of α_j

N	α_1	α_2	α_3
1	11/10	–	–
2	1/10	2/3	–
3	1/2	0	1/3

$$g_i = \langle w_i, g \rangle = \frac{i(3i + 8)}{2(i + 2)(i + 4)} \tag{9.99}$$

Then, the α_j are given by (9.85) and therefore f can be approximated by (9.95) for any fixed N. For successive approximations when $N = 1, 2, 3$, the following matrix equations are respectively formed according to (9.81),

$$\left[\frac{1}{3}\right][\alpha_1] = \left[\frac{11}{30}\right]$$

$$\begin{bmatrix} \frac{1}{3} & \frac{1}{2} \\ \frac{1}{2} & \frac{4}{5} \end{bmatrix} \begin{bmatrix} \alpha_1 \\ \alpha_2 \end{bmatrix} = \begin{bmatrix} \frac{11}{30} \\ \frac{7}{12} \end{bmatrix}$$

$$\begin{bmatrix} \frac{1}{3} & \frac{1}{2} & \frac{3}{5} \\ \frac{1}{2} & \frac{4}{5} & 1 \\ \frac{3}{5} & 1 & \frac{9}{7} \end{bmatrix} \begin{bmatrix} \alpha_1 \\ \alpha_2 \\ \alpha_3 \end{bmatrix} = \begin{bmatrix} \frac{11}{30} \\ \frac{7}{12} \\ \frac{51}{70} \end{bmatrix}$$

whose solution for α_j, as was aforementioned, can be directly obtained from (9.85). Table 9.2 shows successive approximations of α_j as N is increased.

By taking these values into (9.95), the numerical solution is plotted in Figure 9.20, where the analytic solution is also included for comparison purposes. As can be seen, the numerical approach for $N = 3$ is identical to the analytic plot and hence they are exactly overlapped.

9.5   Software Available

The previous section briefly presented the most popular numerical methods developed nowadays. Each of them deserves not only a separate chapter, but a whole book, as can be seen in many of the references given at the end of this chapter. This implies that various details have to be considered if any of them is going to be programmed for a specific problem, some of which have not been addressed since that is out of the scope of this book. However, the fundamental concepts have been introduced and some generalities were explained in order to provide the basic notions. On the other hand, one can

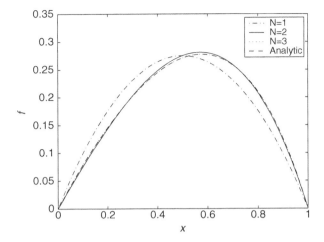

FIGURE 9.20
Successive approximations by the method of moments.

find several works not only on applications of these methods, but also with discussion of software architecture aspects [38–40], open software [27] or some implementation aspects for MATLAB [41, 42]. To date, there is commercial software based on different numerical methods. In what follows, some of the programs are briefly presented to illustrate their potential and the numerical basis on which they operate.

9.5.1 NEC

After the publication of Harrington's book [34], several computer programs were developed at the end of 1960s and the early years of 1970s [43]. As a result of this evolution, and from the focus on shipboard communication applications, the *Numerical Electromagnetic Code* (NEC) was created in 1977 at Lawrence Livermore National Laboratory, USA, as a migration of the named *Antenna Modeling Program* (AMP), which used the Electric Field Integral Equation (EFIE) for wires and the Magnetic Field Integral Equation (MFIE) for surfaces.[9] Thus, the idea was that a broader range of frequency was covered and different structures could be simulated from a few to large wavelengths in size using asymptotic techniques. Then, the first version of NEC (1977) is based on spline expansion for current. The following versions of NEC had the following features [43]: NEC2 (1980) introduced the Sommerfeld integral and interpolation for wires above the ground and the numerical Green's function; NEC3 (1983) improved its capability through the Sommerfeld solution

---

[9]A description of the EFIE and some references to applications of it and the MFIE can be found in [36].

for buried wires and wires penetrating the ground interface; NEC4 (1990) included improved numerical precision for low frequencies, insulated wires and the possibility of varying the radius. Nowadays, the main force of NEC is on wire antennas, which can be simulated related with ground planes so realistic like ships. Basically, NEC uses the MoM as "engine" to solve the integro-differential equations that describe a specific problem, therefore it is usually referred as the "NEC (Method of Moments)."

A very popular and relatively recent variation of NEC is named the *SuperNEC*, which is a software that works with MATLAB, hence it is necessary to have installed it in order that users can interacts with SuperNEC. This program includes a set of most common antennas, which can be modified through some of their parameters to achieve a specific new design. It is also possible to create new structures from MATLAB. Classical antenna examples included are dipoles, monopoles, Yagi, log-periodic, etc., but there also are fractal antennas or built up antenna arrays.

In any case, once the antenna under analysis has been defined, different results can be obtained, like radiation pattern (either two-dimensional or three-dimensional), current distribution on the antenna, impedance plots versus frequency, near electric and magnetic field, efficiency, among others. In order to carry out its calculus, SuperNEC uses EFIE, which is solved by the MoM.

9.5.2   HFSS

The *High Frequency Structural Simulator* (HFSS) developed at Carnegie Mellon University during 1980s is based on the FEM and is currently used not only for antenna design, but also for transmission lines, filters, and so on. According to Ansys, the firm of HFSS, through different solvers this software allows users to select the proper one to analyze the electromagnetic problem at hand. The features of HFSS are[10]:

- EM solver technologies

- HFSS interface

- Circuit simulation extensions

- Advanced finite antenna array simulation

- Automatic adaptive meshing

- Mesh element technology

- High-performance computing

- Advanced broadband SPICE model generation

- Optimization and statistical analysis

---

[10]www.ansys.com

9.5.3 CST STUDIO SUITE

The *Computer Simulation Technology* (CST) STUDIO SUITE [44] is general-purpose software specialized to simulate electromagnetic structures whose numerical core is based on the *Finite Integration Technique* (FIT). As is mentioned in [45], the FIT was developed by Weiland in 1977 as a spatial discretization scheme for Maxwell's equations. In this way, a finite calculation domain is defined, which must naturally include all considerations of the problem to solve. As in any numerical method, one of the essential aspects of the FIT is the creation of the mesh. The small elements that make up the mesh, named *grid cells*, can be hexahedral or tetrahedral grids, as long as all the cells exactly fit each other.

Now, as explained by Demenko et al. in [46], there are equivalents of the FIT and the FEM. In their paper, the authors pointed out that basically the differences between the methods rely on how space is discretized and how equation coefficients are set up.[11] Moreover, in [46] it is shown that FIT equations may be considered a special case of the FEM formulation. Then, provided the equivalents between both methods and the explanation has been given of the FEM, no more details of the FIT are described in this section, but the interested reader can consult [45] and its references.

Regarding the CST STUDIO SUITE, it provides the possibility to simulate 3D structures, giving radiation pattern plots, reflection coefficient (magnitude and phase), current distribution, and transient analysis, among others. Currently, the CST STUDIO SUITE is made up of the following modules [44]:

- CST Microwave Studio

- CST EM Studio

- CST Particle Studio

- CST Cable Studio

- CST PCB Studio

- CST Mphysics Studio

- CST Design Studio

Throughout this book different simulations have been presented, which, as was stated before, were generated using the module CST Microwave Studio. This is because it covers a high frequency range, both in a transient and in a time harmonic state, which is definitely important for UWB simulations. In particular, this module is made up in turn by three solvers, which concern high frequency electromagnetic field problems: *Transient Solver*, *Frequency Solver* and *Eigenmode Solver* (see [44] for more details).

---

[11]The formulation of the FIT discrete space is equivalent to hexahedral FEM elements of eight nodes and 12 edges.

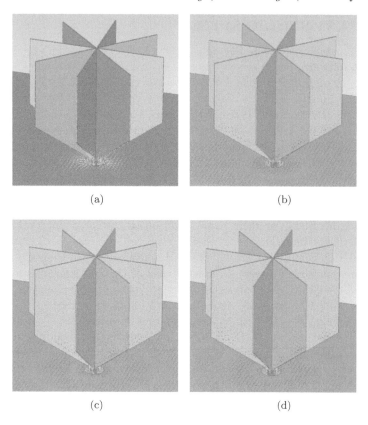

(a)   (b)

(c)   (d)

FIGURE 9.21
Current distribution for the bi-orthogonal UWB antenna: (a) 5 GHz, (b) 10 GHz, (c) 15 GHz, (d) 20 GHz.

Simulation results described in this book were given for the magnitude and phase of the reflection coefficient, whose values indicate, respectively, the impedance matching and phase linearity of the different antenna designs discussed in Chapters 6 and 7. Another important parameter simulated with CST Microwave Studio in these chapters is the radiation pattern. As explained in Chapter 2, from its results it is possible to determine the pattern bandwidth, i.e., the behavior of the radiation pattern through a certain frequency range. These results were essential due to the classification of UWB antennas, omni-directional, and directional, adopted in this text.

In order to illustrate another parameter that can be simulated with CST Microwave Studio, let us take five UWB antenna designs discussed in Chapters 6 and 7. Particularly, current distribution on two omnidirectional antennas (the bi-orthogonal and the rectangular planarized) and three directional antennas (leaf-shape, quasi-Yagi and RPMA) were generated whose results are shown in Figures 9.21–9.25. The parameters and dimensions of these

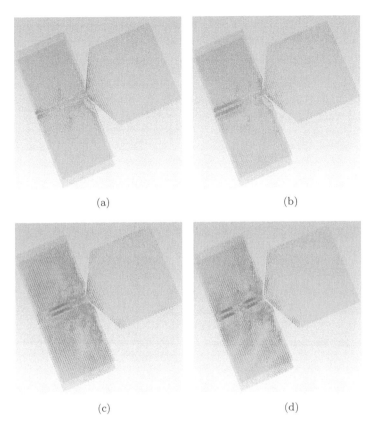

(a) (b)

(c) (d)

FIGURE 9.22
Current distribution for the planarized monopole UWB antenna: (a) 5 GHz,
(b) 7.5 GHz, (c) 10 GHz, (d) 15 GHz.

structures can be found in their corresponding sections. In all cases the Transient Solver was used for different values of *Lines per wavelength.*[12] The boundary conditions in all simulations were *Open add space*, which means that the simulator introduces perfectly matched microwave absorber material at the boundary. This is with the aim to guarantee that boundaries appear to be an open space.

In the first example corresponding to the bi-orthogonal antenna, a resolution of 10 lines per wavelength was considered. The current distribution

---

[12]This parameter sets the spatial resolution (i.e., sampling rate) of the field. A setting of 10, for instance, implies that an EM wave propagating along one of the coordinate axes is sampled at least 10 times.

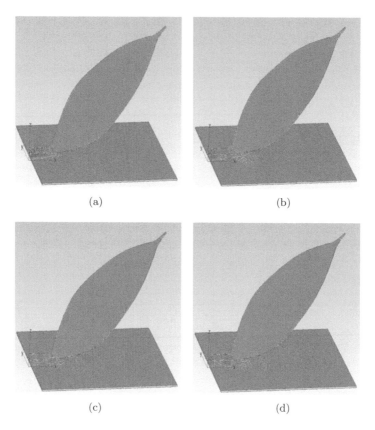

(a)   (b)

(c)   (d)

FIGURE 9.23
Current distribution for the leaf-shaped UWB antenna: (a) 5 GHz, (b) 10 GHz, (c) 15 GHz, (d) 20 GHz.

depicted in Figure 9.21 is mapped to a scale between 0–46.7 A/m. For the planarized monopole UWB antenna, a setting of 15 lines per wavelength was introduced, for which the results shown in Figure 9.22 are given in a scale of 0–77.6 A/m. In the context of directive antennas is the leaf-shaped UWB antenna, whose current distribution is in a scale of 0–25.4 A/m (see Figure 9.23) for a setting of 10 lines per wavelength. The following example of this group is the quasi-Yagi UWB antenna, for which the parameter of lines per wavelength was set to 17. The scale of the current distribution depicted in Figure 9.24 for this antenna is between 0–146 A/m. Finally, Figure 9.25 shows the current distribution of the RPMA, which was simulated using 17 lines per wavelength and whose scale goes from 0–48.8 A/m.

It is important to note the current distribution of the directive antennas, which, as expected, impacts on the radiation pattern. Therefore, depending

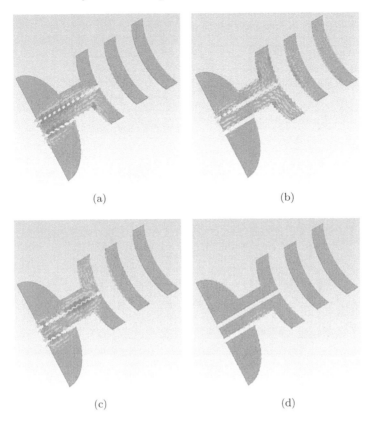

(a)   (b)

(c)   (d)

FIGURE 9.24
Current distribution for the quasi-Yagi UWB antenna: (a) 5 GHz, (b) 10 GHz,
(c) 15 GHz, (d) 20 GHz.

on the variations of this distribution, smooth or abrupt changes in the shape
of this parameter are presented as frequency changes (see Chapter 7). As has
been pointed out in different sections of the book, this behavior has to be evaluated (either by simulations or measurements) for a wide span of frequencies
in all cases of UWB antennas.

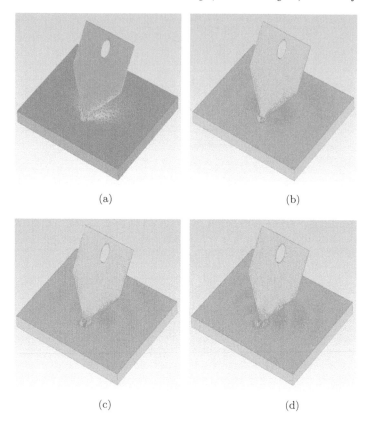

(a)   (b)

(c)   (d)

FIGURE 9.25
Current distribution for the RPMA: (a) 5 GHz, (b) 10 GHz, (c) 15 GHz, (d)
20 GHz.

Bibliography

[1] P. Franklin. *Differential Equations for Electric Engineers.* John Wiley &
    Sons, New York, 1933.

[2] R. P. Agnew. *Differential Equations.* McGraw-Hill, New York, 1942.

[3] W. E. Boyce and R. C. DiPrima. *Elementary Differential Equations and
    Boundary Value Problems.* John Wiley & Sons, New York, 1965.

[4] E. D. Rainville. *Elementary Differential Equations.* Macmillan, New
    York, 5th edition, 1974.

[5] F. Sauvigny. *Partial Differential Equations 1 Foundations and Integral Representations.* Springer, London, 2nd edition, 2012.

[6] R. C. Booton, Jr. *Computational Methods for Electromagnetics and Microwaves.* John Wiley & Sons, New York, 1992.

[7] K. Umashankar and A. Taflove. *Computational Electromagnetics.* Artech House, Norwood, MA, 1993.

[8] J. L. Volakis, A. Chatterjee, and L. C. Kempel. *Finite Element Method for Electromagnetics: Antennas, Microwave Circuit, and Scattering Applications.* Wiley & IEEE Press, 1998.

[9] A. F. Peterson, S. L. Ray, and R. Mittra. *Computational Methods for Electromagnetics.* IEEE Press, New York, 1998.

[10] K. H. Huebner, D. L. Dewhirst, D. E. Smith, and T. G. Byrom. *The Finite Element Method for Engineers.* John Wiley & Sons, New York, 2001.

[11] S. N. Makarov. *Antenna and EM Modeling with MATLAB.* John Wiley & Sons, New York, 2002.

[12] L. M. Magid. *Electromagnetic Fields, Energy, and Waves.* John Wiley & Sons, 1972.

[13] J. W. Arthur. The evolution of the Maxwell's equations from 1862 to the present day. *IEEE Antennas and Propagation Magazine,* 55(3):61–81, June 2013.

[14] A. Sommerfeld. *Eelctrodynamics – Lectures on Theoretical Physics,* volume III. Academic Press Inc., 1952.

[15] R. M. Wilmotte. The distribution of current in a transmitting antenna. *IEE Proceedings of the Wireless Section,* 3(8):136–146, 1928.

[16] L. La Paz and G. A. Miller. Optimum current distribution on vertical antennas. *Proceedings of the IRE,* 31(5):214–232, 1943.

[17] T. Morita. Current distribution on transmitting and receiving antennas. *Proceedings of the IRE,* 38(8):898–904, 1950.

[18] E. Hankui and T. Harada. Estimation of high-frequency current distribution on an antenna. In *1998 IEEE International Symposium on Electromagnetic Compatibility,* volume 2, pages 673–678, Denver, Colorado, USA, 1998. IEEE.

[19] J. Sosa-Pedroza, J. L. López-Bonilla, and V. Barrera-Figueroa. La ecuación generalizada de Pocklington para antenas de alambre de forma arbitraria (in Spanish). *Científica,* 9(2):83–86, 2005.

[20] C. A. Balanis. *Antenna Theory: Analysis and Design*. John Wiley & Sons, 3rd edition, 2005.

[21] A. Cardama-Aznar, L. Jofre-Roca, J. M. Rius, J. Romeu-Robert, and S. Blanch-Boris. *Antenas (In Spanish)*. Alfaomega, 2004.

[22] C. Jordan. *Calculus of Finite Differences*. Chelsea Publishing Company, New York, 1965.

[23] G. Boole. *Calculus of Finite Differences*. Chelsea Publishing Company, New York, 5th edition, 1970.

[24] K. S. Yee. Numerical solution of initial boundary value problems involving Maxwell's equations in isotropic media. *IEEE Transactions on Antennas and Propagation*, AP–14(3):302–307, 1966.

[25] M. F. Iskander, M. D. Morrison, W. C. Datwyler, and M. S. Hamilton. A new course on computational methods in electromagnetics. *IEEE Transactions on Education*, 31(2):101–115, 1988.

[26] S. C. Chapra and R. P. Canale. *Numerical Methods for Engineers*. McGraw-Hill, New York, 2nd edition, 1988.

[27] I. R. Çapoğlu, A. Taflove, and V. Backman. Angora: a free software package for finite-difference time-domain electromagnetic simulation. *IEEE Antennas and Propagation Magazine*, 55(4):80–93, 2013.

[28] R. W. Clough. The finite element method after twenty-five years: a personal view. *Computer & Structures*, 12:361–370, 1980.

[29] J. Jin. *The Finite Element Method in Electromagnetics*. John Wiley & Sons, New York, 2nd edition, 2002.

[30] A. Bondeson, T. Rylander, and P. Ingelström. *Computational Electromagnetics*. Springer, New York, 2005.

[31] R. F. Harrington. Origin and development of the method of moments for field computation. *IEEE Antennas and Propagation Magazine*, 32(3):31–36, June 1990.

[32] L. V. Kantorovich and G. P. Akilov. *Functional Analysis in Normed Spaces*. MacMillan, New York, 1964.

[33] R. F. Harrington. Matrix methods for field problems. *Proceedings of the IEEE*, 55(2):136–146, February 1967.

[34] R. F. Harrington. *Field Computation by Moment Method*. Macmillan, New York, 1968.

[35] L. L. Tsai and C. E. Smith. Moment methods in electromagnetics for undergraduates. *IEEE Transactions on Education*, E–21(1):14–22, February 1978.

[36] S. M. Rao, D. R. Wilton, and A. W. Glisson. Electromagnetic scattering by surfaces of arbitrary shape. *IEEE Transactions on Antennas and Propagation*, AP-30(3):409–418, 1982.

[37] R. F. Harrington. The method of moments – a personal review. In *IEEE Antennas and Propagation Society International Symposium*, volume 3, pages 1639–1640, 2000.

[38] L. Carísio Fernandes and A. J. Martins Soares. Software architecture for the design of electromagnetic simulators. *IEEE Antennas and Propagation Magazine*, 55(1):155–168, 2013.

[39] T. P. Stefański. Electromagnetic problems requiring high-precision computations. *IEEE Antennas and Propagation Magazine*, 55(2):344–353, 2013.

[40] S. Aksoy and M. B. Özakin. A new look at the stability analysis of the finite-difference time-domain method. *IEEE Antennas and Propagation Magazine*, 56(1):293–299, 2014.

[41] S. Makarov. MoM antenna simulations with MATLAB : RWG basis functions. *IEEE Antennas and Propagation Magazine*, 43:100–107, 2001.

[42] G. Toroglğu and L. Sevgi. Finite-difference time domain (FDTD) MATLAB code for first- and second-order EM differential equations. *IEEE Antennas and Propagation Magazine*, 56(2):221–239, 2014.

[43] G. J. Burke, E. K. Miller, and A. J. Poggio. The numerical electromagnetics code (NEC)–a brief history. In *IEEE Antennas and Propagation Society International Symposium*, volume 3, pages 2871–2874. IEEE, 2004.

[44] *CST STUDIO SUITE 2006 Advanced Topics*. www.cst.com, 2006.

[45] M. Clemens and T. Weiland. Discrete electromagnetism with the finite integration technique. *Progress in Electromagnetics Resarch, PIER*, 32:65–87, 2001.

[46] A. Demenko, K. Sykulski, and R. Wojciechowski. On the equivalence of the finite element technique and finite integration formulations. *IEEE Transactions on Magnetics*, 46(8):3169–3172, 2010.

Appendix A

Nabla: The Differential Operator

CONTENTS

For arbitrary a, b, c coordinates, the Nabla differential operator is given by

$$\nabla = \left(\mathbf{i}_a \frac{\partial}{\partial a} + \mathbf{i}_b \frac{\partial}{\partial b} + \mathbf{i}_c \frac{\partial}{\partial c} \right) \tag{A.1}$$

where $\mathbf{i}_a$, $\mathbf{i}_b$, $\mathbf{i}_c$ is the corresponding set of unit vectors. Then,

$$\nabla^2 = \nabla \cdot \nabla = \left(\frac{\partial^2}{\partial a^2} + \frac{\partial^2}{\partial b^2} + \frac{\partial^2}{\partial c^2} \right) \tag{A.2}$$

A.1   Rectangular Coordinates

Let any scalar field be U and the vector field be $\mathbf{A}$. For x, y, z rectangular coordinates with $\mathbf{i}_x$, $\mathbf{i}_y$, $\mathbf{i}_z$ the set of unitary vectors,

$$\nabla U = \mathbf{grad}\ U = \mathbf{i}_x \frac{\partial U}{\partial x} + \mathbf{i}_y \frac{\partial U}{\partial y} + \mathbf{i}_z \frac{\partial U}{\partial z} \tag{A.3}$$

$$\nabla \cdot \mathbf{A} = \mathbf{div}\ \mathbf{A} = \frac{\partial A_x}{\partial x} + \frac{\partial A_y}{\partial y} + \frac{\partial A_z}{\partial z} \tag{A.4}$$

$$\begin{aligned} \nabla \times \mathbf{A} = \mathbf{curl}\ \mathbf{A} \quad &= \quad \mathbf{i}_x \left(\frac{\partial A_z}{\partial y} - \frac{\partial A_y}{\partial z} \right) + \mathbf{i}_y \left(\frac{\partial A_x}{\partial z} - \frac{\partial A_z}{\partial x} \right) \\ &+ \quad \mathbf{i}_z \left(\frac{\partial A_y}{\partial x} - \frac{\partial A_x}{\partial y} \right) \end{aligned} \tag{A.5}$$

$$\nabla^2 U = \mathsf{Lap}\ U = \frac{\partial^2 U}{\partial x^2} + \frac{\partial^2 U}{\partial y^2} + \frac{\partial^2 U}{\partial z^2} \tag{A.6}$$

A.2   Cylindrical Coordinates

Let any scalar field be U and the vector field be $\mathbf{A}$. For r, φ, z cylindrical coordinates with $\mathbf{i}_r$, $\mathbf{i}_\varphi$, $\mathbf{i}_z$ the set of unitary vectors,

$$\nabla U = \mathbf{grad}\ U = \mathbf{i}_r \frac{\partial U}{\partial r} + \mathbf{i}_\theta \frac{\partial U}{\partial \theta} + \mathbf{i}_z \frac{\partial U}{\partial z} \tag{A.7}$$

$$\nabla \cdot \mathbf{A} = \mathbf{div}\ \mathbf{A} = \frac{1}{r} \frac{\partial}{\partial r}(r A_r) + \frac{1}{r} \frac{\partial A_\varphi}{\partial \varphi} + \frac{\partial A_z}{\partial z} \tag{A.8}$$

$$
\begin{aligned}
\nabla \times \mathbf{A} = \mathbf{curl}\ \mathbf{A} \ =\ & \mathbf{i}_r \left(\frac{1}{r} \frac{\partial A_z}{\partial \varphi} - \frac{\partial A_\varphi}{\partial z} \right) + \mathbf{i}_\varphi \left(\frac{\partial A_r}{\partial z} - \frac{\partial A_z}{\partial r} \right) \\
+\ & \mathbf{i}_z \left[\frac{1}{r} \frac{\partial}{\partial r}(r A_\varphi) - \frac{1}{r} \frac{\partial A_r}{\partial \varphi} \right]
\end{aligned}
\tag{A.9}
$$

$$\nabla^2 U = \mathbf{Lap}\ U = \frac{1}{r} \frac{\partial}{\partial r} \left(r \frac{\partial U}{\partial r} \right) + \frac{1}{r^2} \frac{\partial^2 U}{\partial \varphi^2} + \frac{\partial^2 U}{\partial z^2} \tag{A.10}$$

A.3   Spherical Coordinates

Let any scalar field be U and the vector field be $\mathbf{A}$. For r, θ, φ spherical coordinates with $\mathbf{i}_r$, $\mathbf{i}_\theta$, $\mathbf{i}_\varphi$ the set of unitary vectors,

$$\nabla U = \mathbf{grad}\ U = \mathbf{i}_r \frac{\partial U}{\partial r} + \mathbf{i}_\theta \frac{1}{r} \frac{\partial U}{\partial \theta} + \mathbf{i}_\varphi \frac{1}{r \sin \theta} \frac{\partial U}{\partial \varphi} \tag{A.11}$$

$$
\begin{aligned}
\nabla \cdot \mathbf{A} = \mathbf{div}\ \mathbf{A} \ =\ & \frac{1}{r^2} \frac{\partial}{\partial r} \left(r^2 A_r \right) + \frac{1}{r \sin \theta} \frac{\partial}{\partial \theta} \left(\sin \theta A_\theta \right) \\
+\ & \frac{1}{r \sin \theta} \frac{\partial A_\varphi}{\partial \varphi}
\end{aligned}
\tag{A.12}
$$

$$
\begin{aligned}
\nabla \times \mathbf{A} = \mathbf{curl}\ \mathbf{A} \ =\ & \mathbf{i}_r \left[\frac{1}{r \sin \theta} \frac{\partial}{\partial \theta} \left(\sin \theta A_\varphi \right) - \frac{1}{r \sin \theta} \frac{\partial A_\theta}{\partial \varphi} \right] \\
+\ & \mathbf{i}_\theta \left[\frac{1}{r \sin \theta} \frac{\partial A_r}{\partial \varphi} - \frac{1}{r} \frac{\partial}{\partial r} \left(r A_\varphi \right) \right] \\
+\ & \mathbf{i}_\varphi \left[\frac{1}{r} \frac{\partial}{\partial r} \left(r A_\theta \right) - \frac{1}{r} \frac{\partial A_r}{\partial \theta} \right]
\end{aligned}
\tag{A.13}
$$

$$\nabla^2 U = \text{Lap } U \quad = \quad \frac{1}{r^2} \frac{\partial}{\partial r} \left(r^2 \frac{\partial U}{\partial r} \right) + \frac{1}{r^2 \sin \theta} \frac{\partial}{\partial \theta} \left(\sin \theta \frac{\partial U}{\partial \theta} \right)$$

$$+ \quad \frac{1}{r^2 \sin^2 \theta} \frac{\partial^2 U}{\partial \varphi^2} \tag{A.14}$$

Appendix B

Some Concepts Related to Differential Equations

CONTENTS

B.1   Generalities

Differential equations are used to model the variation of one quantity as a function of another one (or more than one). When the differential equation presents one or more variables which depend on only one parameter, it is said it is an *ordinary differential equation*, otherwise we are addressing *partial differential equations*. For example,

$$m\frac{\mathrm{d}^2 x}{\mathrm{d}t^2} = F \tag{B.1}$$

$$\frac{\partial^2 u}{\partial x^2} + \frac{\partial^2 u}{\partial y^2} = 0 \tag{B.2}$$

In the case of Equation (B.1), Newton's law for the position $x(t)$ of a particle acted on by a force F, we can see how there is a variation of the distance, x, as a function of time, t (i.e., a velocity), and there is not any other independent variable. Hence this equation is an example of an ordinary differential equation.

On the other hand, Equation (B.2), corresponding to the Laplace's or the potential equation, presents the potential u as a function of two spatial variables, which means that $u = u(x, y)$ and therefore it is a partial differential equation.

B.2   Order

The order of an ordinary or partial differential equation is the order of the highest derivative that appears in the equation. Thus, for example Equation (B.1) is a second order ordinary equation.

B.3   Differential Equation Systems

When a problem is described by more than one differential equation (either ordinary or partial and of any order), it is said that we have a system of differential equations. A classical example is the set of partial differential equations formulated by Maxwell and presented in Chapter 9:

$$\nabla \times \mathbf{E} + \mu_0 \frac{\partial \mathbf{H}}{\partial t} = 0 \tag{B.3}$$

$$\nabla \times \mathbf{H} - \varepsilon_0 \frac{\partial \mathbf{E}}{\partial t} = \mathbf{J} \tag{B.4}$$

$$\nabla \cdot \varepsilon_0 \mathbf{E} = \rho \tag{B.5}$$

$$\nabla \cdot \mu_0 \mathbf{H} = 0 \tag{B.6}$$

where $\mathbf{E}$ represents the electrical field, $\mathbf{H}$ the magnetic field, ε_0 is the permittivity of vacuum, and μ_0 is the permeability of vacuum. The source of electromagnetic energy is given by the density of current variant with time ($\mathbf{J}$), which is associated to a charge density also variant with time (ρ). Finally, ∇ is the differential operator.

B.4   Initial and Boundary Value Conditions

In the field of differential equations, conditions over which problems are defined are the key. Depending on the physical nature of the problem, the solution of a differential equation can be subject to certain conditions. There are two possible conditions: initial and boundary value conditions. To explain the former, let us assume the first order ordinary equation given by

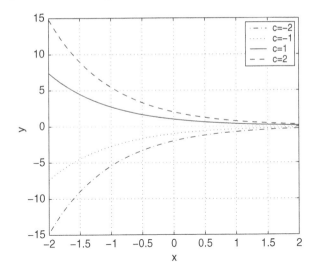

FIGURE B.1
Family of curves for Equation (B.8) when $a = 1$.

$$\frac{dy}{dx} + ay = 0 \tag{B.7}$$

where a is a real constant. A possible solution for (B.7) could be

$$y = ce^{-ax} \tag{B.8}$$

with c an arbitrary constant and therefore actually Equation (B.8) represents infinitely many solutions of (B.7). Equation (B.8) can be geometrically plotted as shown in Figure B.1 for $a = 1$. As can be seen we have a family of curves parametrized on c, which is named *integral curves* of (B.7).

Hence each integral curve is the geometric representation of the corresponding solution of the differential equation. If a particular solution is specified at a single point, let us say (x_0, y_0), then

$$y(x_0) = y_0 \tag{B.9}$$

is referred to as an *initial condition*, and together with Equation (B.8) forms what is known as an *initial value problem*. It is worth mentioning that the term initial conditions usually refers to an initial time, i.e., time is the independent variable.

In contrast, when the differential equation that describes a certain problem in a region R is subject to several conditions over the boundary of R, these are called *boundary value conditions*, and therefore there is a *boundary value problem*. These conditions describe a specific physical condition at the boundary of the body under analysis. In the case of the Maxwell's equation systems, for a thin layer surface between two regions R_1 and R_2 the boundary conditions are:

$$\mathbf{n} \times (\mathbf{E}_1 - \mathbf{E}_2) = 0 \tag{B.10}$$

$$\mathbf{n} \times (\mathbf{H}_1 - \mathbf{H}_2) = \mathbf{K} \tag{B.11}$$

$$\mathbf{n} \cdot \varepsilon_0 (\mathbf{E}_1 - \mathbf{E}_2) = \rho_s \tag{B.12}$$

$$\mathbf{n} \cdot \mu_0 (\mathbf{H}_1 - \mathbf{H}_2) = 0 \tag{B.13}$$

$$\mathbf{n} \cdot (\mathbf{J}_1 - \mathbf{J}_2) + \nabla_S \cdot \mathbf{K} = -\frac{\partial \rho_s}{\partial t} \tag{B.14}$$

which relate the electric and magnetic fields both within a structure and to its surroundings. For (B.10)–(B.14) $\mathbf{n}$ represents the component normal to the surface, $\mathbf{K}$ the surface current density, and subscripts 1 and 2 are for the two regions.

B.5   Existence and Uniqueness

The first question that arises is if the differential equation has a solution. Then we are talking about the *existence* of the solution of that equation. This aspect comes from the mathematical formulation of a physical problem. If it is properly formulated as a differential equation, a solution should exist.

On the other hand, there exists the question of *uniqueness* of the solution of a differential equation, or, in other words, is the solution found unique or there could be others? As can be seen, this question is related to the initial or boundary value conditions aforementioned (for example if for Equation (B.7) the initial condition would be $y(0) = 2$, a particular solution is given by substituting $x = 0$ and $y = 2$ in (B.8), in such a way that $c = 2$ and then it is possible to determine a unique solution). Thus, there should be a theorem that mathematically enunciates the response to this question. The above is known as the *Existence and Uniqueness Theorem* and can be enunciated as follows, in terms of initial conditions,

---

Existence and Uniqueness Theorem

If $f(x, y)$ and $\partial f(x, y)/\partial y$ are continuous functions of x and y in a region $|x - x_0| \leq a$, $|y - y_0| \leq b$ then there is one and only one function $y = y(x)$, defined in some interval $|x - x_0| \leq h \leq a$, which satisfies the differential equation

$$\frac{dy}{dx} = f(x, y)$$

and also the initial condition

$$y(x_0) = y_0$$

Naturally, this theorem can be also defined for a boundary value problem, and to be so specific for the problem at hand.

Appendix C

Poisson and Laplace Equations

CONTENTS

C.1 Poisson's Equation

Let us assume an arbitrary stationary static electric charge distribution known in a limited region of finite volume V' only, whose charge distribution is given by either volume, surface, line or point charge densities. From the basic static electric field laws, it is known that the static $\mathbf{E}$ field must be conservative everywhere, which means

$$\nabla \times \mathbf{E} = 0 \tag{C.1}$$

and that its divergence in the presence of volume charge density, ρ, is given at each point by

$$\nabla \cdot \mathbf{E} = \frac{\rho}{\varepsilon_0} \tag{C.2}$$

with ε_0 the permittivity of vacuum. Due to the conservative nature of the static electric field as is dictated in Equation (C.1), it can always be expressed in terms of the gradient of a scalar electric potential Φ, in such a way that,

$$\mathbf{E} = -\nabla \Phi \tag{C.3}$$

Thus, by substituting Equation (C.3) into Equation (C.2), results as

$$\nabla^2 \Phi = -\frac{\rho}{\varepsilon_0} \tag{C.4}$$

Equation (C.4) is known as *Poisson's equation.*

C.2   Laplace's Equation

If $\rho = 0$ in (C.4), we have

$$\nabla^2 \Phi = 0 \qquad\qquad (\text{C.5})$$

Equation (C.5) is known as *Laplace's equation*.

Index